Dup
594432
FCBSADCL-1

92 MARSHALL, J. Howard (James
Marshall), 1905-
Howard), 1905-
Done in oil : an
autobiography /

Copyright © 1994 by J. Howard Marshall II
Manufactured in the United States of America
All rights reserved
First edition

Frontispiece: J. Howard Marshall II; photo by Barbara Sutro

The paper used in this book meets the minimum requirements
of the American National Standard for Permanence
of Paper for Printed Library Materials, Z39.48-1984.
Binding materials have been chosen for durability.

Library of Congress Cataloging-in-Publication Data

Marshall, J. Howard (James Howard), 1905–
 Done in oil : an autobiography / by J. Howard Marshall II ;
edited with an introduction by Robert L. Bradley, Jr.
  p. cm.
 Includes bibliographical references and index.
 ISBN 0-89096-533-1
 1. Marshall, J. Howard (James Howard), 1905– ,
2. Businessmen—United States—Biography. 3. Petroleum industry and trade—United States—History—20th
century. I. Bradley, Robert L., 1955– . II. Title.
HD9570.M34A3 1994
338.7'6223382'092—dc20
[B]   94-10098
                                                    CIP

IN MEMORY OF
Bettye Bohanon Marshall
AND
Ralph K. Davies

Contents

List of Illustrations   *page*   ix
Preface   xi
Author's Acknowledgments   xvii
Editor's Acknowledgments   xxi
The Oil Trilogy of J. Howard Marshall II: Editor's Introduction   3

CHAPTER

1 Beginnings   12
2 Regulating Oil at Interior, 1933–35   31
3 High Adventure with Standard of California, 1935–41   67
4 Back with Government: War Petroleum Planning, 1941–44   113
5 Building Ashland Oil & Refining, 1944–51   156
6 Producing at Signal Oil & Gas Company, 1952–60   197
7 Creating Momentum at Union Texas Petroleum, 1961–67   231
8 On My Own, 1968–Present   246

Final Letter to Ralph K. Davies   269
Selected Readings   271
Index   273

Illustrations

Marshall as a lad	page	13
Captain of tennis team		17
National Guardsman checking East Texas well valves		29
"36 Section" of Elk Hills field		32
Bypass valve in East Texas field		43
W. F. "Big Fish" Fisher		48
Wells spudding in Wilmington field		83
Signal's yacht		96
Standard's presidents		108
Harold Ickes		117
Big Inch pipeline		129
Transportation meeting		133
Refinery		136
PAW dinner honoring Ralph Davies		147
Ashland headquarters		161
Marshall and Blazer		173
Garth Young, Essat Gaffar, and Jim Brooks		191
Sam Mosher		198
Signal Hill		200
Great Northern field—before and after		203
Lake Arthur South field		233

Illustrations

Staff of Marshall Petroleum 251
Koch Industries board meeting 255
Oscar Wyatt with Marshall and E. O. Buck 257
Bettye Bohanon Marshall 262

Preface

For many years I have threatened to write a book about the oil industry. In my library rests an autographed copy of Max Ball's *This Fascinating Oil Business*. Max's note to me reads, "To Howard, who could write a better book if he would." Once, years ago, I was tempted to try. But a senior law partner with Pillsbury, Madison & Sutro reminded me of the old adage, "Would that mine enemy would write a book." Since I have made some enemies in my life, I restrained myself—at least until now.

The book I did finally begin to write, entitled *Ten Years of California Oil*, sought to contrast what the news media reported about California oil with what I thought was really occurring. When I looked at my notes one evening, I suddenly realized that I knew, or believed I knew, what was really going on in California because my clients and associates in the industry trusted me enough to keep me so informed. Since honest lawyers may not, and decent men do not, tell tales involving those who trust them, I tossed my notes in the fireplace one cold winter evening in San Francisco.[1]

---

1. Editor's note: Marshall did write a brief synopsis of California's oil and gas conservation law through 1937. See J. Howard Marshall, "Legal History of Conservation of Oil and Gas in California," in Section of Mineral Law, American Bar Association, *Legal History of Conservation of Oil and Gas: A Symposium* (Baltimore: Lord Baltimore Press, 1939), 28–36.

PREFACE

The same contrast between truth and fiction arose much later when I was asked to comment on a manuscript that was later published as *A History of the Petroleum Administration for War*.[2] It was a good recitation of petroleum planning during World War II. The book was mostly factual, interesting, "true" (in a general sense), but, as I suspect with most historical writing, a story of how it *should* have happened, not exactly how it *did* happen. As one of those who helped conceive, create, and administer the Petroleum Administration for War (or PAW) and who served as its chief counsel and assistant deputy, I thought I really knew how it happened. But, once again, out of respect for the confidences of so many of my associates, I decided it was best to leave history as it had already been written. Even now, despite the passing of time and so many old associates, I am not certain what limitations I should place on myself in telling various untold strategies, including those involved in the formation and administration of the PAW. To whatever extent I find myself still unable to tell a complete story, however, I invite the reader to look between the lines.

It would be easy and personally amusing merely to write an autobiography of my own checkered career. But most autobiographies, consciously or unconsciously, end up portraying the author as some sort of hero. About one such bit of partly imaginative corporate history, one reader remarked: "The hero should have been called 'Saint Paul' rather than just plain Paul, because in hundreds of pages he failed to make a mistake."

Of course it seldom makes much sense for anyone to write a book merely about himself. Even if, for the moment, it makes good light entertainment, such a book and its author will soon be gone and forgotten with the rest. Unless the writing portrays something greater than the writer's life, it is hardly worth the effort to write it. For myself, I shall describe the oil industry as experienced by an individual who has lived, worked, competed, and played in it ever since the early 1930s.

Now, as almost always during the past, the titular heads of the oil business have generally been inarticulate when it comes to writing about their own profession. Those few who really understood the

---

2. John W. Frey and H. Chandler Ide, *A History of the Petroleum Administration for War: 1941–1945* (Washington: Government Printing Office, 1946).

industry were, by and large, doers rather than writers—people of action, not words—risk-takers who dominated and directed the allocation of their companies' assets, not corporate bureaucrats seeking to preserve what their predecessors had accumulated.

One of the old oil giants, Al Weil, a great lawyer who left his practice to head General Petroleum in California, was once threatened, along with most of the rest of us in the California oil industry, with an anti-trust indictment. The then-head of Union Oil of California met Al on the street and asked whether he thought we might be sent to jail. Al replied that if he went it would give him a chance to write his memoirs. When asked who would read them, Al answered, "You will, you old SOB, to see what I say about you." Perhaps only those who fear what I may incidentally say about them will ever read this book, but those who might have the most to fear are no longer present.

For those who may still be present, a story about my own fears of some years ago may allay some concerns. A decade after I had left the presidency of Ashland Oil and Refining Company, at a time when I was serving as an executive vice-president of Allied Chemical, I had a call from Rex Blazer, then chairman of Ashland. He told me that his uncle, Paul Blazer, founder of Ashland, had commissioned Otto Scott to write the history of Ashland. Scott had asked if he could talk with me. I asked Rex if this was all right with his Uncle Paul. Rex laughed and replied "Regretfully, yes." I asked Rex to tell Paul (then near death) that there was nothing I would relate to the author except things that would reflect credit on Paul Blazer. I had learned too much from Paul about many aspects of the oil business I had never known before ever to want to tell not only the truth but the whole truth. Rex called me the next day to tell me he had relayed my message and been told by Paul to say he was grateful.

Since I am a curious person, before Scott interrogated me, I asked him a simple question—why did he want to talk to me? He answered that he had spent many hours with Paul Blazer. Suddenly something had struck him as strange. He told Blazer that the years when Ashland broke out from being merely a strong independent refiner on the banks of the Big Sandy River in Kentucky to a major company corresponded exactly with the period when I had served as president of Ashland. Yet Blazer had never mentioned my name. Scott told Blazer, "I want to talk to this man." I asked about Blazer's reply: "Well, I suppose you have to."

PREFACE

Incidentally, when Scott's book was published,[3] I hurriedly checked the index under Marshall to see how badly I was cut up. I was not, and I hope my old friends will feel the same way if they ever bother to read this book.

As with all matters of life and living, what has happened will never happen exactly that way again. History is ever not quite a circle. The Petroleum Industry, less static than most, I would argue, will never exactly recreate its own past. Yet, since "past is prologue," and the people in the industry today are shaped by its past, recounting that past here may help explain and perhaps predict how the industry is likely to deal with the future.

When I first inadvertently stumbled into an interest in oil, I operated under certain assumptions concerning the oil industry's reputation as a monopoly. It had been run by some fellow named Rockefeller—a ruthless individual (or, at least, portrayed as such by Ida Tarbell in two volumes about the history of Standard Oil).[4] Stuart Chase said the industry was wasteful beyond belief.[5] My own college yearbook, in its 1926 caricature section, has me standing on a soapbox, shirttail out, delivering a tirade against "The Standard Oil Company." Years later, as a partner in a law firm that served as general counsel for Standard of California, I sometimes wondered if I would ever have to explain that picture.

Much, much later I was to learn that Chase was right—the waste was frightful and included billions of barrels of oil left irrecoverable in the ground, hundreds of millions of barrels allowed to run down creeks or dumped into earthen pits, trillions of feet of natural gas flared and blown into the air, tens of thousands of unnecessary wells drilled in a race to get claims to the same pool of oil and gas, and hundreds of inefficient refineries built merely to dump crude oil into the market faster than other producers drawing from the same pool. So many thousands of so-called service stations multiplied like locusts along the highways and on almost every metropolitan street corner that someone once remarked that the only reason there were only four service stations at an intersection was that there were only four corners.

---

3. Otto Scott, *The Exception* (New York: McGraw-Hill, 1968).
4. Ida Tarbell, *The History of the Standard Oil Company*, 2 vols. (New York: McClure, Phillips, and Company, 1904).
5. Stuart Chase, *The Tragedy of Waste* (New York: Macmillan Company, 1928).

PREFACE

What Stuart Chase partly missed—and what Ida Tarbell never understood at all—was that this wasteful mess resulted not from monopoly but from the excesses of competition. Another author, John Flynn, recognized this later.[6] Here he showed old John D. as a product of his time, seeking to bring what he regarded as order out of the chaos of competition, which characterized the oil business almost from its inception.

My late wife, an industry veteran herself, claimed that I was the last of a vanishing breed of oilmen—perhaps, perhaps not. But I might be one of the last, in this age of specialization, who has worked and directed operations in nearly every phase of the oil industry— from exploration to marketing, from the courtroom to government, and from the major corporations to the independents. I once was persuaded by the then-president of the Continental Oil Company, L. F. McCollum, to make a speech. He introduced me by reading a biography from *Who's Who in America* and concluded with: "And now ladies and gentlemen I want to introduce a man who doesn't seem to be able to hold a job." I could only answer by confessing that although it was true, perhaps it was better than a ten-year service pin from Continental Oil (now Conoco).

Once I boasted, too vainly I fear, that although there were many who knew far more than I about a particular phase of the oil business, I felt capable of matching wits with anyone about the way the different pieces fit together. Today we speak of oil as an integrated industry, and we complain that our troubles arise because in a modern corporate or governmental bureaucracy, there are so few who can think in integrated terms. One's own department, one's own little empire, one's own organization chart, one's own political position in either the government or the company—these circumscribe the thinking of too many people.

If this book is worthwhile, it will be only because it presents the oil industry in integrated terms—as it is and not as either the public or writers imagine. Decades ago I drafted an address for Ralph Davies, then a vice-president of Standard of California, which stated that the economics of oil obeyed the laws of hydraulics—for we never, or so rarely ever, see any oil or its major products. It moves in a continuous

---

6. John Flynn, *God's Gold: The Story of Rockefeller and His Times* (New York: Harcourt, Brace & Company, 1932).

stream, out of sight and under pressure from the wells, through the pipelines and refineries, through other lines or transports to filling stations, homes, and industry. Raise, lower, or alter the supply or demand at any point in this closed system, and the rest of the system is immediately affected. This theory is simple, you say—yes, but how many understand it? Perhaps this book—through both its lighter and more serious aspects—will provide concrete examples of how that system operates as a whole.

One warning before I start. This story will not be garnished with endnotes (except by the editor). The best footnote ever written, in my opinion, appeared accidentally in one of Thurman Arnold's books. He had apparently assigned the task of researching a footnote to an assistant in the Department of Justice. The footnote read, "Find some authority to support this damn statement." The error was caught before many copies of the book appeared on the street. For myself I shall rely on a lawyer's memory—imperfect, certainly, but accurate enough not to exaggerate or improve the story too much in the telling.

Author's Acknowledgments

My boyhood and early intellectual influences are mentioned in chapter 1. My primary influences in the industry—all contemporaries—were Ralph Davies, J. R. Parten, Al Weil, and Sam Mosher. If they were only successful oilmen, which these four certainly were, they would have been just friends. But each had an intellectual side—a longing not only to "do" but to understand the "why" of doing—which made them inspirational friends. I've run across too many successful oilmen who lacked even a modicum of this quality, persons whom I've referred to in my weak moments as "stuffed shirts."

Ralph Davies was my closest friend and mentor. Actually, we were each other's mentors. California's oil problems brought us together in 1933, when I was with the Interior Department and Ralph was with Standard of California. When I left Interior in 1935, I joined Ralph's company, California Standard, due in part to his recommendation and persistence in wanting me on-board. We immediately became best friends and got on the fast track together. Ralph came up through the ranks in marketing, while I was more familiar with exploration and production. Back then he was an oilman and I a lawyer, but before long we each wore both hats. We very nearly got the top two jobs at

Standard (now Chevron) five years later, but fate sent us in another direction, as chapters 3 and 4 describe.

Despite his setback at Standard, Ralph had a great career as an independent. Historians should regard him as one of the best oilmen of his generation; he was certainly the best I ever met. The reminiscences of Ralph published in *Ralph K. Davies: As We Knew Him* (San Francisco: Weiss Printing Company, 1976), as well as my own in this book, are a testament to his personal character and business acumen. My farewell letter to Ralph appears at the end of the book.

J. R. Parten, long-time chairman of Woodley Petroleum, was the strongest independent oilman I ever met. He was the proverbial Texan sporting a large physical stature and emotional toughness. There was nothing "false" or synthetic about J. R. He was a leader of his generation—generation 4 in terms of the industry lineage beginning with John D. Rockefeller—and a driving force behind his chosen causes. As in the case of Davies, a book on J. R.'s life would prove valuable to posterity.

Al Weil had a profound influence on my life in the oil industry. Like me, he was an attorney, oilman, and independent who later headed General Petroleum, a subsidiary of Socony-Vacuum (now Mobil). In addition to being a smart, practical oilman, he was also an intellectual, although he might balk at such a characterization. In Al's case, however, this extra dimension resulted from neither an abundance of education nor a passion for reading. Rather, his knowledge resulted from asking "why," as did both Ralph's and J. R.'s. Al's greatest mistake, for which he never forgave himself, was reversing his board vote to remove Ralph Davies from Standard's presidency, a story related in chapter 3.

A fourth major influence on my career was Samuel Mosher, founder and long-time head of Signal Oil and Gas Company (now Allied-Signal Company). Sam had beautiful oil judgment, which he put to good use in creating the largest independent oil company on the West Coast. I got to know Sam well when Ralph Davies and I negotiated gasoline contracts between Standard and Signal. Sam reentered my life in a big way when he visited me in an Ashland hospital in late 1951 and persuaded me to join Signal. I was at a crossroads in my career (having left Ashland Oil and Refining), and Sam's offer launched another important phase of my career, as chapter 6 describes.

Of the next generation of oilmen, two stand out and are described

AUTHOR'S ACKNOWLEDGMENTS

in chapter 8. One is Charles Koch, the architect of Koch Industries, Inc., who, like Ralph Davies, can claim to be one of the best of his generation. Charles is a motivating force not only with his company but with his "classical liberal" political philosophy, as well.

Oscar Wyatt of the Coastal Corporation is another oilman of Koch's generation who has captured my imagination. He towers above most of his industry peers—which may help to explain his popularity problems. Oscar would have fit in very well in the industry era that preceded him.

A note concerning the book itself: it is about my oil career, not my personal life. But I should mention two of the fairer influences. My first wife, Eleanor Pierce, the mother of my two sons, was my college sweetheart, who never really understood my driving passion for the oil business. Few could, perhaps. While my work was my mistress, Eleanor was as faithful a wife as I could ever have wanted. My second wife, Bettye Bohanon, was a rare woman, whose professional career I describe in chapter 8. She was instrumental in the writing of the book.

Finally, I wish to acknowledge the editorial efforts of Robert L. Bradley, Jr., whose sincere interest in the history of the American petroleum industry has benefited the manuscript greatly. It is largely through his persistence that my scribblings on a yellow legal pad have turned into a handsome book.

—J. Howard Marshall II

Editor's Acknowledgments

Preparing this book for publication has been a rare privilege. It is not often that one gets to befriend a legend, and J. Howard Marshall II is certainly that.

The comments and efforts of the following individuals are greatly appreciated: Eyvonne Scurlock of Marshall Petroleum, who coordinated the efforts between the editor and author; Pierce Marshall, who carefully read the manuscript and offered numerous suggestions and insights to shape the final draft; George Pearson, formerly of Koch Industries, Inc., and Austin O'Toole of the Coastal Corporation, who provided factual information on their companies for the final chapter; David LeBeau of the Chevron corporate library and Michael Rogan of the Conservation Committee of California Oil Producers, who tracked down facts and figures from the California Standard years; Elliott Powers and Jay Grubb, who sharpened chapter 7; Melvin Payne, Stuart Johnston, and Louis Moore, who added facts for the opening sections of the final chapter, and Robert L. Bradley, who reviewed the page proofs and made final corrections.

—Robert L. Bradley, Jr.

Done in Oil

The Oil Trilogy of J. Howard Marshall II: Editor's Introduction

ROBERT L. BRADLEY, JR.

The professional career of J. Howard Marshall II has had three distinct yet interrelated phases, constituting an adventure trilogy. He began his career as an intellectual and, with the publication of this book, will leave the same way. In between he has been one of the most important petroleum regulators of this century and a distinguished petroleum industry executive at both the integrated and nonintegrated levels.

The manuscript had gathered dust for over a decade when Marshall let me read it—I was the first person to read it in its entirety. I was eager to read his accounts of some famous government-industry episodes—particularly the East Texas hot oil war, which occurred from 1933 to 1935, the market-demand proration debate in California later in the same decade, and petroleum planning during World War II.

As I read the manuscript, I realized that not only did it richly supplement the historical record, but it was also a frank, highly readable account of a multifaceted career. In terms of longevity, accomplishments, variety, perspective, and "high adventure," Marshall's career rivals all but one—that of old John D. Rockefeller himself. Consequently, *Done in Oil* should rank as a classic autobiography.

Career

A thumbnail sketch of Marshall's life and trilogy in the oil business is necessary to put the book in perspective. Born on January 24, 1905, in Philadelphia, Pennsylvania, Howard distinguished himself in both academics and athletics. He was a promising tennis player, who as a teenager received praise and tutelage from Bill Tilden himself. He also excelled at soccer as a two-time All-American goaltender for intercollegiate-champion Haverford College in the mid-1920s. On the scholastic side, he was captain of the debate team and editor of the school paper at both George School and Haverford.

Marshall began the first part of his trilogy in oil in academia. After receiving his AB from Haverford in 1926, he entered Yale Law School the next fall. As a top student, he was chosen as case editor for the *Yale Law Journal*. On one submission dealing with the legality of a state law prorating oil production to "market demand," Marshall and Norman Meyers, a Ph.D. economist, investigated the legalities of unrestrained oil output from the East Texas and Oklahoma City fields. At issue here was whether the "rule of capture," a legal doctrine assigning first title of oil production to physical possession, could be modified by state regulation to limit output and prevent physical waste—thus increasing prices for the industry. The result of this excursion was the lead article for the *YLJ* of November 1931. Marshall and Meyers coauthored the article, entitled "Legal Planning of Petroleum Production."

That same year Marshall received the Juris Doctor degree magna cum laude from Yale and accepted an offer to stay there as assistant professor and assistant to the dean. Still interested in the oil proration question, he and Meyers in February, 1933, published in the *YLJ* an analysis entitled "Legal Planning of Petroleum Production: Two Years of Proration," an updated version of their earlier article. The article caught the attention of the secretary of the Interior, Harold Ickes, who at the time was grappling with the problems of regulating petroleum "overproduction." Ickes persuaded Marshall and Meyers to join Interior as assistant solicitors for the summer of 1933.

The "Doctor and the Dean," as they were nicknamed, spent two riotous years wrestling with Interior's "oil problem." They wrote a "Code of Fair Competition for the Petroleum Industry," which was approved by FDR in August, 1933, pursuant to the National Industrial

Recovery Act (or NRA). They set up the government's Petroleum Administration Board and the advisory industry group, the Planning and Coordinating Committee, to promote industry stability amid rampant wellhead overproduction as part of the New Deal's effort to improve the profitability of American business.

During this period Marshall discovered the advantages and disadvantages of regulation. Setting comprehensive floor prices for crude oil and petroleum products, Ickes's first choice, was rejected by Marshall and Myers as unworkable. They also viewed marketing codes of fair competition to prevent the flood of crude oil from turning into a surplus of gasoline as "hell on wheels." Their answer was to limit supply at the source, the wellhead. That meant prorating the great fields of Texas, Oklahoma, and California, in particular. With "hot oil" producers—and lawyers such as Marshall's adversary, F. W. "Big Fish" Fischer—outfoxing the Texas Railroad Commission, a federal solution was necessary. Marshall's innovation, which finally achieved production control (and thus price stability for the industry), was a tender system whereby each oil producer had to document that each barrel of crude had been legally produced under state conservation law before it could be transported in interstate commerce. The legal anchor of the tender system was Section 9(c) of the NRA, which prohibited the interstate movement of illegally produced oil. When Section 9(c) was declared unconstitutional by the Supreme Court in early 1935, Marshall, under a forty-eight-hour deadline from Sen. Tom Connally, authored a remake—the Interstate Transportation of Petroleum Products Act, better known as the Connally Hot Oil Act.

With the federal tender system rescued to complement state enforcement efforts, the East Texas and Oklahoma City fields came under control. California, however, lacked a proration law. It was in this state (not coincidentally) that Marshall would soon find himself in demand by the very industry he had helped to regulate.

With the invalidation of the NRA, Marshall found himself at a crossroads. The thirty-year-old fast-tracker had three career alternatives: staying on at Interior as assistant secretary behind Ickes, rejoining the Yale Law School faculty (where he had been technically on a leave of absence since June, 1933) as assistant professor with an inside track to becoming dean, or joining Standard of California in a loosely defined managerial role. Longing to enter the oil fray from the industry side, he signed up with Standard in June, 1935. There he joined up with his newfound friend and alter ego, Ralph K. Davies, and spent the next five years in "hard work and high adventure."

Standard's problem, and that of the other major integrated companies (or "majors"), was not a shortage but a surplus of oil. Without state proration, independents producing wide-open in new fields threatened the fragile price structure as well as the downstream investments (i.e., refining and marketing) predicated on it. Five episodes in Marshall's Standard period—two with the company and three with its law firm—revolved around controlling crude production and gasoline marketing on the major's terms. Proving his skill as a lawyer, Marshall wrote several chapters of legal history, which included creating an informal gasoline purchase network, formulating a much copied production-sharing contract between a municipality and an industry firm, authoring the California State Lands Act of 1938, winning the Bolsa Chica Gun Club lease, and spearheading support for the Atkinson Proration Bill of 1939.

It was the defeat of the proration bill that eliminated the frontrunner for Standard's presidency, Ralph Davies, and took Marshall, who would have been second in command under Davies, out of the company's legal mainstream. But as European tensions grew in the 1930s, the two sought to organize a planning bureaucracy, not unlike the U.S. Fuel Administration in World War I and the Petroleum Administration Board during the New Deal, to help run the domestic oil industry. The idea for the Office of Petroleum Coordinator (later the Petroleum Administration for War) was born at a San Francisco restaurant and featured Harold Ickes, still the secretary of Interior, as the political point man. Under coordinator Ickes, Davies and Marshall ran the show as deputy coordinator and chief counsel respectively.

For the next four years, the OPC/PAW regulated all aspects of the petroleum industry except for price. Elusive wellhead conservation regulation in California and Illinois was now possible for the first time since the demise of the New Deal in 1935. Industry-wide coordination, in place of "cutthroat competition," was brought about. Two huge federal oil pipelines—the Big Inch and the Little Big Inch—and many smaller ones were constructed. Refineries were required to pool their know-how, standardize their civilian products, and yield huge quantities of 100-octane gasoline during World War II. The defining characteristic of the planning effort was primary reliance on the private sector and a largely cordial industry-government partnership. In the 1970s, in contrast, pervasive federal regulation of the oil industry was adversarial. Marshall's account powerfully suggests that in a second best world of regulation, the cooperative approach with industry-expert staffing is preferable to what has been the rule in recent decades. To this extent the efforts of Marshall and Davies helped to win the war.

While Marshall can be criticized for regulating per se, his philosophy was to avoid regulation whenever possible. He recognized that baiting the hook with higher prices or tax incentives was preferable to what today is called command-and-control. He also blocked an eleventh-hour attempt by Abe Fortas, under-secretary of Interior, to nationalize the East Texas and Wilmington (California) oil fields, the two largest in the country.

It is interesting that Marshall's rather untainted description of the World War II planning exercise has not been contradicted to date. Market economists have questioned the need for wartime and emergency planning, but no scandals or smoking guns have been documented within the OPC/PAW's industry-oriented planning effort.

Tired of wartime planning, Marshall moved back to industry in 1944. This time the call came from Paul Blazer, the brilliant, though egocentric, founder and chairman of Ashland Oil and Refining Company. Just as Marshall was getting settled at his new company, a phone call to Blazer from Harry Truman sidetracked Ashland's new president, making him general counsel of the Allied Commission on Reparations. This would be Marshall's last detour into government service. After that, the third part of Marshall's trilogy—petroleum industry executive—would be played out.

The Ashland era for Marshall was bittersweet. On the one hand, Marshall honed his downstream skills at the feet of a master and played an important role in leading Ashland out of the independent ranks into a near-major status. On the other hand, Marshall tells us, Blazer became increasingly difficult to work with, which led to an acrimonious split in early 1952. But it would always be the Ashland years—seven long years of dedicated and intense effort—that Marshall saw as the highlight of his entrepreneurial career. A slightly different fate would have had Marshall, with his knowledge of exploration and production, succeeding an ailing Blazer as chairman and, perhaps, leading the company into the ranks of a bonafide major.

As Marshall entered the third decade of his oil trilogy, he had a lot of experience but little to show for it financially. The disappointment of the two near misses with California Standard and Ashland weighed heavily upon him. It was time, in his words, "to work for Marshall." The right offer came from Signal Oil and Gas, which was like Ashland a strong independent. Sam Mosher, Signal's chairman, offered Marshall an executive position with the understanding that two-thirds of Marshall's time was for Signal and the last third was for Marshall.

The Signal era (1952–60) would be professionally and personally rewarding. Sam Mosher did not have the same kind of personality as Blazer and gave Marshall plenty of room to operate. For Signal's share of his time, Marshall won a lucrative, long-lived oil concession in Argentina by pioneering a production-sharing contract with the national oil company that improved upon the standard 50-50 production agreement, while meeting Argentina's constitutional prohibition against foreign ownership of crude reserves. Several major production plays in the Los Angeles Basin were developed. The biggest success, however, was for Marshall's own account.

While president of Ashland, Marshall and others had conceived of building a Midwest refinery to process Canadian crude oil, but Blazer was not interested in his subordinate's idea. Now with Signal, Marshall and J. R. Parten resurrected the idea to construct a refinery in St. Paul, Minnesota, that would process twenty-five thousand barrels a day. Marshall gambled by becoming a stockholder, but his Great Northern Refining Company, completed in 1955, became very successful. Almost from the beginning, the refinery was profitable, and fifteen years later, Marshall's Great Northern interest would evolve into a position in Koch Industries, Inc., which would increase substantially in value.

The next stop in Marshall's career was Union Texas Petroleum (1961–67). Parting with Mosher was difficult, but a lucrative financial package (and the opportunity to be captain of the ship rather than first mate) proved incentive enough.

The first order of business was to integrate Union Texas with two recent acquisitions and reform a natural gas contract that imperiled the health of the company. Later, Union Texas was itself bought by Allied Chemical to become an integrated oil, gas, and petrochemical company. In the process Marshall found himself a first mate after all, so, he decided, it was again time to "work for Marshall." After an unsuccessful refining venture, the rest of his career would revolve around exploration and production through private placement limited partnerships and Marshall Petroleum, Inc.

Associations and Directorships

In his postwar years, Marshall accumulated some associations and directorships of note, including: trustee, National Petroleum Association; founding member, National Petroleum Council (1946–78); president, National Stripper Well Association (1946); director, American Petroleum Institute (1945–present); director, Texas Commerce Bank (1957–82); director, Koch Industries, Inc. (1969–present); and director, Coastal Corporation (1973–present). Of these, his two most interesting and consequential associations were the last two.

The growth of privately held Koch Industries, Inc., represents a model for corporate America. Under the direction of Charles Koch, who made a rare exception to take Marshall in as an outside partner, Koch Industries has quietly become, over several decades, one of the largest and most successful privately held firms in the United States. Koch's uniquely entrepreneurial approach, however, was challenged in the early 1980s, when two of the four brother-owners attempted to take the company public. Marshall's family interest proved to be the balance of power in the struggle to keep the company private, which is one of the keys to Koch's continuing success and competitive advantage.

The Coastal Corporation was a second company that was unusually prosperous during Marshall's association with it. His interpretation of Coastal's controversial founder and longtime chairman, Oscar Wyatt, deviates from the generally accepted one. Having seen similar criticism of other successful industry executives in his long career, Marshall identifies many of Wyatt's critics as less talented competitors. Marshall's interpretation is supported by the fact that Coastal prospered during periods when most of its peer companies did not.

The Evolution of a Regulator

An underlying theme of this book is the evolution of Marshall's views regarding the necessity and desirability of government intervention in the petroleum industry. In his first published article on the subject, he advocated government price-fixing to achieve industry stability. When Marshall was asked to engineer the real thing several years later, he quickly recognized that real-world complexity made this approach folly. Respect for the market is evident in his repeated references to the problems created by the Department of Energy's heavy-handed regulation in the 1970s. Marshall also recognized that the downstream cooperative agreements (such as pooling agreements to buy surplus gasoline) were self-help mechanisms for companies that were not monopolistic but, rather, fragile alliances to offset the perennial problems associated with rampant competition.

Two reservations stood in the way of Marshall's favoring total market reliance. The first was his belief that the rule of capture, a property rights assignment awarding first title to crude oil to physical possession, required government-enforced wellhead proration to balance supply and demand in the short run and preserve supply in the long run. The market-oriented economist has several rebuttals to this idea. First, another private-property rights assignment could have awarded the discoverer of a contiguous oil reservoir a much bigger claim to simplify the "transactions cost" problem of multiple-ownership under the rule of capture. Second, given the nature of the rule of capture, a variety of government interventions—from antitrust statutes to conservation law itself—hampered entrepreneurial solutions to overdrilling and overproduction.[1] Professor Walton Hamilton, a Yale Law School professor specializing in economics (and one of Marshall's early influences), was correct when he asserted that oil production was unique because of the supply implications of reservoir mechanics. But the "market failure" spawned by the rule of capture was not necessarily greater than the "government failure" of mandatory conservation law.

The second reservation was Marshall's almost automatic assumption that central government planning was necessary in place of "blind competition" for an efficient industry war-mobilization effort. However, greater reliance on "baiting the hook," to use Marshall's term, could have reduced government involvement in industrial projects. The Big Inch and Little Big Inch pipelines, for example, could have

---

1. For an elaboration of these points, see my *Oil, Gas, and Government: The U.S. Experience* (Lanham: University Press of America, 1995), chapters 2–4.

been privately constructed with the right throughput agreements and tax incentives. Unregulated prices in place of price ceilings could have solved the what, when, and where problems associated with petroleum supplies and infrastructure materials. Without planning, greater interfirm rivalry could have incited greater entrepreneurial discovery to benefit both the civilian and wartime theaters with a more efficient use of resources.

During this time there was virtually no intellectual or reasoned industry second-guessing of market-demand proration and wartime planning. J. Howard Pew (president of Sun Oil), for example, the penultimate industry critic of government regulation in the first half of the century, lobbied for governmental oil proration when Sun's competitive fortunes were threatened. Questioning the rationale for wartime planning risked being perceived as unpatriotic. From today's perspective, on the other hand, earlier government-industry episodes can be critically evaluated.

A letter dated March 22, 1979, from Marshall to then Energy Secretary James Schlesinger is illuminating. It was advice from the century's most successful petroleum regulator to its least successful (if there can be such a thing as a "successful" regulator). Marshall's stints during the New Deal and World War II contrast sharply with Schlesinger's controversial tenure as the first energy czar at the Department of Energy. Whereas Schlesinger fell into the morass of price and allocation regulation, resulting in the gasoline lines of 1979, Marshall foresaw the problems of price-setting and talked his superior out of it. Marshall's tenure has stood the test of time, but Schlesinger's era has been criticized as one of the greatest regulatory failures in U.S. history.

In his introduction Marshall states that a good autobiography must relate to something larger than the author. This book does. It fills in important gaps in the business and regulatory history of the U.S. petroleum industry, and it offers the present generation of oil executives an insightful look at the tumultuous era that preceded theirs. The book is wonderfully written—it was completed in one long-hand draft, for the most part—by a man who in 1994, at age ninety, has not lost the booming courtroom voice or the infectious glimmer in his eye. J. Howard Marshall II's six decades of contributions to the petroleum industry are now for posterity.

CHAPTER ONE

Beginnings

I was born on January 24, 1905, in Germantown, Pennsylvania, a suburb of Philadelphia, to Samuel Furman and Annabelle (Thompson) Marshall. My birth came upon the second anniversary of the death of my grandfather, J. Howard Marshall, for whom I was named. I had a delightful old Quaker grandmother who, together with my mother, probably influenced my younger life most.

J. Howard Marshall I was an independent steelman who helped put together a combination of blast furnaces and rolling mills that he and his brother later sold to Andrew Carnegie as part of the formation of U.S. Steel. On the deal they received around eighteen million dollars in bonds and common stock. That was tremendous money, especially in 1902 dollars, and my grandfather told my grandmother to sell the U.S. Steel common when he died, which she did.

Many years later when I was checking up on some estate questions, I found out how much the U.S. common was worth. I asked my mother, "Does thee know what thee would be worth today if thee had kept that U.S. Steel?" She said she had no idea, and I told her she would be worth about one hundred million dollars. But this assessment was made in the Roaring Twenties; after October, 1929, its value was far less.

Marshall as a young lad.

By the time my grandmother and aunts, who had lived on this inheritance for many years, died, the total value had shrunk to around one million dollars. It had been invested by Philadelphia investment bankers, who managed to lose it by being too conservative, and the Great Depression had almost wiped my grandmother out. She wanted to give the last remaining half-million to me, but I refused the bequest, as I believed my mother would have wanted. I asked her to give it to my children to see what they could do with it. By that time I did not need it, anyway—but that is getting ahead of the story.

My father, Furman Marshall, was a gifted engineer who graduated from the Drexel Institute in Philadelphia. Unfortunately, he became a misplaced engineer when the family sent him to help run Marshall Brothers, a steel jobbing company. He disliked it and failed in a managerial role. He did not have the toughness to be a good businessman. My mother had that.

Dad and I were great companions, though. We built a thirty-two-foot sloop on our lawn by hand over a three-year period. When my brother and I sold it many years later, it was like selling a piece of myself.

My brother, John T. Marshall, inherited my father's engineering skills. At the age of twelve, he was able to take a Swiss watch apart, put it back together, and make it work.

I used to be asked endlessly in law school whether I was related to John Marshall, who served as chief justice of the Supreme Court from 1801 until his death in 1835. At the time I didn't know. As it turns out, however, I'm descended from one of his brothers, and, like all Marshalls, I inherited one of his major characteristics—big ears. Perhaps their size is appropriate, however, because I've learned to listen with them. As a young lawyer I was always talking, but I've learned that sometimes it's better to keep your mouth shut and your ears open.

When I was twelve, I was stricken with a serious case of typhoid fever. I've always said I must have been "born to be hanged" because, during my illness, I had a temperature in excess of 109 degrees for fifteen minutes and should have died. When I came back to life, I was left with what the doctors call a typhoid hip. The inflammation had virtually destroyed my left hip joint and about five inches of the bone from the ball and the socket. The doctors said I would never walk again. My mother, a very strong person, disagreed, and, after I'd been on crutches many months she took them away, burned them, and told me to walk. I staggered, stumbled, and fell—but the more I used my hip, the less it hurt.

One day I tried riding my brother's bicycle. Strangely enough, the more I rode that bicycle, the better I got. Although there was no real power in my left leg, it had to follow the motion of the pedal. Though I didn't know it at the time, I'd created my own type of physical therapy, and it must have helped reform the joint.

By the time I went to George School, a Quaker preparatory school outside Philadelphia, I had recovered sufficiently to play soccer. This was a particularly tough sport for a young man with a bad left hip, but in certain ways my injury proved to be an advantage. I had become quite overdeveloped in my right side; thus, I could kick a soccer ball with that leg a good fifty or sixty yards. I became a good prep school player and participated in the Philadelphia soccer league.

About this time I took a job caretaking the tennis courts at Buck Hill Falls in Philadelphia, where my grandmother had a cottage. I received a dollar per day for my work at this old Quaker resort. But I had an additional privilege—the use of the tennis courts where Bill Tilden, the leading tennis player of his generation, would regularly practice.

I was practicing one day when Bill first noticed me. He told me I had good reflexes and that, if I would work at it, he would make me into a tennis player in spite of my hip injury. That was a great challenge for any youngster. So I worked at it intensely—ten and twelve hours a day—as Bill wanted. (One of the chapters in Tilden's *Match Play and the Spin of the Ball*, published in 1925, was entitled "The Value of Intensive Practice.") I got to where I couldn't hit a ball incorrectly if I wanted to. By the time I got to college, I had become a pretty good tennis player.

Bill taught me not only court strategy but also how to hit the ball hard, accurately, and consistently. He had what was known in those days as a cannon-ball serve. He was one of the first big serve-and-volley people, but he had beautiful ground strokes, and I tried my best to emulate him.

When I played, "gamesmanship" was sometimes necessary. My limp was always obvious, but I exaggerated it in the warm up to make my opponent overconfident. I knew I'd developed a reputation when I overheard a rival's coach say, "Look out for that guy with a limp."

I was reminded of Tilden's influence some years later when I played in the Pennsylvania Doubles Championship, which my partner and I won in five long sets. As I walked off the court somebody stopped me to ask, "Mr. Marshall, where did you get that backhand? I haven't seen a backhand like that since Bill Tilden." I looked at him, grinned, and said, "You called it. That's exactly where I learned it."

One of the other things Bill taught me was to never run around the ball. "Learn to hit it where it is," he said. "You are a half a step slow with that leg of yours. That won't hurt you against a player you can beat, but against a good player that will beat you. You must learn to hit the ball hard and get to the net where your reflexes will win for you." I tried to do it just that way and had some tournament championships to show for it.

At George School (1917–22), I was the captain and top player on the tennis team, editor-in-chief of the school paper, president of the Forum, an honorary society, the captain of one of the two debating teams, and a second-team goalie in soccer.

After George School, I went to Haverford College. I was supposed to go to Swarthmore because my family were Hicksite Quakers. Haverford was the Orthodox school. I once asked my grandmother what the difference between the two was, and she told me that on the facing bench in a Quaker meeting, the Hicksites turn their thumbs one way and the Orthodox the other, and that was about it. Later, the two sects finally merged.

The reason I went to Haverford is that I wanted to play soccer. Later I often kidded the president of Haverford by saying that the only reason I enrolled was because I got an athletic scholarship, which, in those days, was not a claim to fame for students. The idea of an athletic scholarship implied an academic shortcoming. In my case this was partly right because, of all things, I flunked the college board examinations in English, despite the fact that I had never earned less than an A in my life in the subject. Somebody got hold of my examination paper, discovered that it was marked 92, and graded it down to a 69. I had misspelled thirteen words. I couldn't spell then, and I can't spell now, though I've never found it much of a hindrance. Haverford admitted me anyway after noticing that I was an all-scholastic goal tender in the Philadelphia soccer league. My timing was perfect: I was entering Haverford the year after their goal tender graduated.

In those days Haverford played all the top teams in soccer—Harvard, Yale, Columbia, Penn, Princeton, Navy, and Army. We were

Marshall (*top center*) as captain of the undefeated George School tennis team. *Courtesy George School*

a school of four hundred, and the reason we were so good was that we drew from two Philadelphia prep schools that did not play football, but groomed their best athletes for soccer. One was Westtown, an Orthodox school, and the other was George School, the Hicksite School where I had attended. There were about a half-dozen Westtown boys and several from George School on my team at Haverford.

We won the intercollegiate championship in my sophomore year at Haverford by beating Penn on Franklin Field. As the match drew to a close, the score was two to one. In the last two minutes, I was lucky enough to stop a penalty kick by getting about five inches of my right hand on the ball and tipping it beyond the post. In the locker room after the game, a doctor who had known my family when I was a child came in and confronted me: "You get off that soccer field. And don't ever get back on because if anyone hits hard, nobody will ever put that hip joint together again." Either the doctor was wrong, which I suspect, or I never got hit from the right angle.

I played three full years of intercollegiate soccer without a substitution. When I returned for my team's fifty-year reunion, David Breum, a sports writer for the *New York Times*, came up to me and asked what it was like playing Swarthmore. I replied arrogantly that they weren't good enough, and in soccer we played them with our junior varsity. I still remember his response: "Well, I'll be damned." Actually, in my era, we usually won against other competition but never played Swarthmore. Later Haverford did play Swarthmore in soccer and often got beat.

We did, however, play them in tennis while I was there. I played four years at Haverford and was captain of the team. In my junior and senior years I never lost, until the last match of my college tennis career, when I was beaten in three long sets by the Swarthmore player. That loss was enough to lose the dual match and that proved to be the anticlimax of my college tennis career.

I graduated from Haverford in the spring of 1926, but kept on playing tennis. My family was afraid I would become a tennis bum, since I regularly traveled on the amateur circuit, though I never went anywhere unless somebody paid my expenses. I never had to spend any money. On the other hand, there was no prize money to pocket—just my ranking to play for. I got all my rackets free from the same company that made Bill Tilden's. All in all, I thoroughly enjoyed my brief stint during and immediately after college as a tournament tennis player.

Years later, when I was practicing law, I had to appear in court on an important case involving millions of dollars. The importance of the case had so unnerved me that I would get sick at my stomach. They way I handled my nerves was to remind myself how I'd felt years before when I went out on the field to play an intercollegiate championship game. I had always been sick at my stomach—until I got my hand on the ball. Then the nausea would go away, and my adrenalin would make me feel even better than normal. Well, it worked in the courtroom just as it had on the soccer field.

Other extracurricular activities at Haverford helped me in my law career as well. Editing the college paper improved my reading and writing skills. Debating was tailor-made for a future trial lawyer. My undergraduate major in chemistry and physics, in addition to the geology I learned from Professor Rowe (whom I'll discuss shortly) after leaving Haverford, allowed me to understand the technical side of the petroleum industry almost as well as the business side. As it turned out, all my college activities served a purpose in later life.

A First Impression of Oil

It has sometimes been asked how an old Philadelphia Quaker ever got involved in a high-risk, speculative, rough-and-tumble business like oil. My conservative Republican family believed it was almost immoral to borrow money. Certainly, few oil men ever succeeded without borrowing a lot of it. Surely not even the most sophisticated computer could have been programmed to predict that I would end up as one of those "wicked oil men" who are said to "bleed the public" and gather "obscene profits." My story's a lengthy one. Like most of life, it resulted from a series of accidents.

My career began in the fall of 1926. After graduating from Haverford, I took a job as a newspaper editor and teacher of freshman history, economics, and English with the first around-the-world "floating university," which took five hundred college students and a university faculty on a voyage around the globe for some eight months. I was qualified to edit the daily paper published aboard the chartered vessel. I had two years' experience on the city desk of the *Philadelphia Enquirer*, writing largely on sporting events, and I knew how a newspaper worked, so my teaching credentials consisted largely of being able to read faster than most of my students.

On board ship I was completely occupied teaching, editing, and assisting in the cruise office. On shore, my time was my own. I soon joined the geology classes of Dr. Rowe—whose first name I never knew and today cannot locate—a professor on leave from Princeton. How fortunate I was to have him as a professor! Later I could claim that, although I had never had a formal course in geology, I could teach one. There could hardly be a better way to learn elementary geology quickly than from a great teacher on a long cruise.

My first brush with the oil industry in operation came during the early weeks of my cruise. As we rode the red cars of the Pacific Electric from San Pedro to Los Angeles, I noticed a forest of derricks that dotted the recently discovered oil fields of the Los Angeles basin. Signal Hill, Domingez, Huntington Beach, Sante Fe Springs—all had wells so closely spaced that sometimes the derricks' legs interlocked. I remember wondering what idiot had designed wells to be drilled in such crazy patterns. A few nights later I looked out across the Los Angeles basin from the top of Mount Wilson (before we ever heard of smog). Miles away in the distance, I could see a flaming torch—a wild well on fire in the Sante Fe Springs field.

Little did I realize then that, in a few short years, I would find myself intimately associated with the California oil industry and the Los Angeles basin. Still less did I anticipate that I would meet in that same basin a young woman in the oil business who would one day become my wife.

It is worth a short digression at this point to contrast my early perceptions of the industry with my later knowledge of it. Years later, after far more detailed study and experience than Dr. Rowe could provide, I came to understand the whys and wherefores of town-lot drilling, wild wells, wasteful practices, and dog-eat-dog competition in every phase of the oil business. That wild well in Sante Fe Springs—was this an example of how a great field blew and burned enough natural gas to supply the city of Los Angeles for several decades? The forest of derricks—did this illustrate one of the important reasons why even from "depleted" oil fields, we generally have left some 75 percent of all the oil originally in the ground? Did the newly discovered oil fields of the Los Angeles basin once again initiate the vicious cycle of feast and famine and horrendous competitive excesses that had heretofore and would hereafter mark the oil business?

Although I did not understand it at the time, what I saw was another episode in a whole series of wasteful chapters in the economic history of the oil industry, a history dating back to the beginning of the business in the middle of the nineteenth century. Each new discovery would inevitably be rushed into production, over-drilled, and inefficiently produced—developers were driven by the "law of capture," which engendered competitive instincts no one could or did control.

The law of capture was defined by Robert Hardwicke as follows: "The owner of a tract of land acquires title to the oil or gas which he produces from wells drilled therein, although it may be proved that part of such oil or gas migrated from adjoining lands."[1] Prices of oil would break to a few cents a barrel in the wild scramble to get a share of the oil out of a common pool before the competition did. There was not time to gather, collect, and use the gas that came with the oil, so it was ruthlessly blown into the air or burned at the wellhead. If one leaseholder complained about a neighboring leasee, the holder's only recourse, said the courts, was to: "Go thou and do likewise." Once a field or a series of fields had blown their heads off, prices would skyrocket, only to have the same old process of feast and famine repeat itself with the next series of discoveries.

Again and again, we witnessed the paradox of predicted long run shortage with short-term surplus. I once said that I started believing in conservation because I was sure we were going to run out of oil. Much later in my career, I began to fear the prescence of too much oil. I've spent most of my business life in oil, wrestling with too much rather than too little. Although the industry is regularly accused of monopoly, it is, if anything, too fond of competition—it has produced too much oil too quickly and sold it too cheaply. As some industry wisecracker once remarked, "Our sales department always sells at less than our cost of production—we lose money on every gallon but make up for it by volume."

Although I did not know it at the time, in the mid-1920s I had caught California in the middle of its feast-and-famine cycle. At the end of World War I, California oil production had drastically declined, and preparations were being made to import and store Mexican crude. Standard of California built enormous concrete reservoirs to receive and store the foreign crude at its El Segundo refinery on the coast near Los Angeles. The great Los Angeles–Basin fields then existed only in the minds of a few imaginative geologists. Indeed, a vice-president of Standard of California said he would drink all the oil that could ever be produced from Signal Hill.

---

1. Robert Hardwicke, "The Rule of Capture and Its Implications as Applied to Oil and Gas," *Texas Law Review* (June, 1935): 409.

As it turned out, he would have had a lot to drink! Shell drilled the discovery well on Signal Hill in 1921, and the industry was off to the races. The discovery proved to be a giant one and was followed quickly by other giants that produced billions of barrels. Ironically, underneath those concrete reservoirs which Standard built to hold imported oil, lay far more oil than they were ever designed to hold, but it took wildcatters, who looked where the oil wasn't supposed to be, to find it.

In the late 1930s, the El Segundo refinery site was offset with a producing well. Standard had to back off and drill a directional well to get under its own tanks. It believed it had drilled a dry hole, but the driller used up the last drilling bit after being told to plug and abandon. The well came in barefoot (without pipe in the hole), for some two thousand barrels a day, right under the refinery. This merely proved what we would learn many times over: oil is where you find it.

The unexpected, unpredictable character of the next series of discoveries illustrates one of the fundamental reasons why no one company or group of companies has ever been able to monopolize this business. They would if they could. Various have been the attempts—universal, the failures. Just when it is thought that oil is controlled and order assured, somebody who doesn't "know any better" upsets the applecart with a new discovery—a major field, a new geologic basin, or an important technical breakthrough.

Doom-and-gloom oil supply forecasts by the U.S. Geological Survey, the Bureau of Mines, the Federal Oil Conservation Board, and a multitude of other experts have all been wrong. More recently, efforts were made to supplement crude oil and natural gas with synthetics. On the demand side, mandatory conservation became the new salvation. With a relaxation of price regulation concerning both natural gas and crude oil, supply became more plentiful. Producers once again began to fear for their financial future. Indeed, by 1982, we witnessed the spectacle of independent gas producers in Western Oklahoma arguing in Washington for a continuance of price controls on shallow gas to hold up the price on deep gas, which was decontrolled and priced out of the market. What is more, independents in favor of "free enterprise" made this argument without blushing! The oil tariff advocates of several years later once again revealed the existence of too much, rather than too little, oil.

BEGINNINGS
A Taste of Business

To return to the main current of this story: after my first around-the-world student cruise, I came to the conclusion that, despite a first-class liberal arts education from Haverford College, I really hadn't learned how to *do* anything. A number of my friends had long ago predicted that I would end up as a lawyer—probably because I led my debating teams in both high school and college and loved to argue. I had always resisted the idea, but now, more from desperation than conviction, I took my savings from the cruise and entered the Yale School of Law. I considered Harvard but decided against it because Harvard impressed me as a mass production system. I picked Yale because of its smaller classes.

By the following summer (1928), I again needed funds. Another floating university-world cruise for students was organized in New York. Because of my previous experience, I found a summer job with the cruise's organizers negotiating steamship contracts and travel arrangements and assembling a college faculty from those willing to devote their sabbatical year to teaching in return for a free trip around the world.

When fall arrived, the promoters of the cruise decided that the man they had selected to manage the cruise en route was incompetent and offered me the job. The five-thousand-dollar salary and all-expense-paid position were more than I could resist. I journeyed to New Haven to ask Bob Hutchins, then the dean of the Yale School of Law, if I would still have my place as a competitor for a position on the *Yale Law Journal*—a position then regarded as the key to getting started in the profession after graduation—if I took a year off. He assured me that my A average was assurance enough but advised me not to take the cruise job, offering me a scholarship as an incentive to stay. He was afraid I wouldn't return to my studies if I left, and he was nearly right.

I dropped out of law school for a year and ran the eight-month around-the-world cruise, learning many things which helped me later in the oil industry. I always credited what I learned about Middle Eastern trading to the able instruction of my assistant on board ship, Consintine Raisies. He taught me how to increase my commission by 5 to 10 percent by assuring hotel managers that the ship would be returning on future cruises.

On board ship the books were all kept by hand. I didn't know bookkeeping when I left, but my father had almost been a C.P.A. at one time. I asked him to teach me double-entry bookkeeping in one week. He replied that learning it so quickly was impossible, but he started me out, and by the time I got across the Pacific two months later, I had learned it.

I used to carry the accounts in two big tin boxes, which were always in my possession. When I drove from one port to the next, I found there were some good tricks to making a profit. One was to go to all the local banks and buy up all the hard currency for the next country on our itinerary because of the slight discount. Then I charged the students not what I paid for it but what the published rates were.

We also realized that students did not need to buy American Express checks when they could deposit their money with the cruise and draw checks in the desired currency at the published rates. We could beat the published rates by 5 or 10 percent to turn the small profit needed for the cruise to return home solvent. I didn't charge the students any premium. I charged them what they would have had to pay had they gone to the bank and changed money. That, fortunately, was not what I paid.

We had another plan for using the students' money. The idea came to me on a New York subway on my way to work one morning. I read that the call money rate in New York (this was the time of the old stock market boom) was about 12 percent. I thought that if I could get the students to deposit their money with the cruise, given that it takes two months for a draft to get from Singapore to New York, we could be drawing the call money rate for the whole time until they cashed their checks. This is the same principle under which travelers' checks operate today. My employer, a New York investment banker, thought this was a stroke of genius.

One other idea also worked to our advantage. I got myself commissioned as a travel agent so I could sell the students insurance on their baggage. There is a tremendous commission—some 20 to 25 percent or more—that made us money. It was pocket-money schemes like these that got us home.

Dean Hutchins was almost right about how taking one year off from law school might keep me from returning. When I returned from my second cruise, Thomas Cook and Sons, a leading travel agency, offered me a job at twenty thousand dollars a year, a huge sum of money for any man my age, to run the student cruises. Somehow I had sense enough to turn it down. This was another one of my lucky accidents, for the year was 1929, and the great stock market crash took the bloom off the travel business for some years to come. I survived pretty well, however, as I had my five thousand from the past year's work, a year of priceless practical experience, and a berth at Yale to ride out the depression that none of us foresaw.

Discovering Oil at Yale

I quickly got back into the flow of law school. I won a place on the editorial board of the *Yale Law Journal,* and, even better, I became its case editor. A case editor's job includes reading the advance sheets looking for material for notes and comments on recent decisions coming through the courts. This new job, strangely enough, led me directly into the oil business, as one of these decisions from a Federal Circuit Court of Appeals involved the constitutionality of the so-called oil proration laws of Oklahoma.

In the 1931 case, *Champlin Refining Company v. Oklahoma Corporation Commission,* Champlin challenged the validity of a 1915 Oklahoma statute authorizing the Corporation Commission to limit (or "prorate") the oil produced from various oil and gas fields in the state. The case raised the interrelated questions regarding the physical and economic effects of regulating the rates of production of individual wells drawing from commonly held underground oil and gas reservoirs. Some argued, as some still do today, that proration was only a thinly disguised scheme to fix high prices. Others took the position that it was necessary to limit production to reasonable market demand, both to prevent physical waste in the fields and to protect the correlative rights of individual owners of wells producing from the same reservoir. Thus were the battle lines drawn in the early 1930s. These arguments were as sterile as the age-old question about which came first, the chicken or the egg. Many in the industry were moved by price; the conservationists were moved by the necessity of limiting the law of capture if we were to get the most oil at the lowest long-run cost. In fact, the two objectives were inseparable.

Both the Circuit Court and, later, the Supreme Court upheld the Oklahoma statute in written opinions, which demonstrated a scanty understanding on their parts of either economics or petroleum engineering. As often happens, the courts were right for the wrong reason. I assigned the Oklahoma case as a case note to a young Oklahoman. His note made no more factual sense than the court's opinion. I put it aside with the thought that some time, when we were short of material, I would spend an evening or two in the library to see what I could do with the note. That time was not long in coming, and I recruited the help of Norman Meyers, a classmate with a doctoral degree in economics. After many nights' work, we concluded the case warranted more than a mere case note, or even a longer comment. It deserved a leading article, which usually would've been written by learned professors rather than mere students. Since neither of us had any compunction about rushing in where we maybe shouldn't have, we wrote the article. We prepared a careful draft over some months and, having read it over, decided we needed to visit the field and check our theoretical conclusions against actual operations.

Both of us were about to graduate. Norman planned to join the staff of the Federal Power Commission in the fall, and I was about to become an instructor and assistant to the dean of the Yale Law School. But between graduation and employment, we had a free summer. So we raised the magnificent sum of five hundred dollars—half of it from the Law School and half from the Yale School of Economics—to finance a long trip through the oil fields of the Southwest to prepare a monograph, which we entitled "Legal Planning of Petroleum Production." We filled the back of my roadster with books, packed our tentative manuscript, and, together with my new wife, set out for the Southwest.

With all the gall you might expect from a couple of arrogant young intellectuals, we descended on anyone who would talk to us: oil and gas lawyers, geologists, petroleum engineers, economists, independents and major companies, even the then-governor of Oklahoma, William "Alfalfa Bill" Murray. He had just declared martial law (on August 4, 1931) and sent in the National Guard to regulate the chaotic and wasteful production emanating from the recently opened Oklahoma City field. We saw the beginning of the great East Texas boom, where Gov. Ross Sterling of Texas followed Murray's precedent on an even larger scale. On August 17, Sterling declared martial law and sent twelve hundred Texas National Guardsmen to halt production of the greatest oil field ever discovered in the contiguous United States.

We watched and learned. We stayed in cheap motels and tourist camps, slept in sleeping bags, and bought "cut" price gasoline for the car. We came home much richer in knowledge. We scrapped our original draft of the article and spent three weeks on the living room floor of my family's home outside Philadelphia rewriting it. Strange as it may seem, we wrote each and every sentence as a team. My mother always remembered the shouting and arguments that punctuated every idea and each sentence. The monograph was finally published that fall in the *Yale Law Journal*.[2] This leading article turned out to be a beginning and not an end for Norman and me.

We both were educated and challenged by a great teacher of law and economics, Dr. Walton Hale Hamilton, a law school professor without a law degree. He taught such courses as the public control of business and constitutional law. "Hammy" was no classical economist. He maintained, rightly I think, that there were no universal laws of economics of steel, coal, oil, textiles, or what have you. Each industry was different and responded to its own peculiar forces. None, except perhaps small agriculture or retailing, followed what he referred to as Adam Smith's economics of petty trade. He taught us that if things went bad in the coal market, an owner could shut down a mine and still have his coal. But as long as the law of capture prevailed, no oil operator could shut down his wells and come back later to produce. The oil would be drained away by those who continued to produce. He also taught us that, whereas in many businesses the highest-cost production comes with the start up, the lowest-cost production with oil comes in the flush period of a new discovery. Then the stuff flows or can be pumped at a high rate at little out-of-pocket cost; higher costs per barrel follow as the wells decline and more and more water must be lifted with the oil. Only where one owner owns an entire field (rare in this country but common in countries where the government owns everything) can an oil field be turned off or on in response to price and the demands of the market.

Over the next two years on the faculty of the Law School, I learned, as do all professors, that you are expected to write and publish. Since that was the route to follow for academic advancement, what could be better than to follow the road on which Norman and I

---

2. J. Howard Marshall and Norman L. Meyers, "Legal Planning of Petroleum Production," *Yale Law Journal* (Nov., 1931): 33–68.

had embarked? So we followed the industry, got together over weekends and vacations, and, two years later, produced a sequel to our first leading article. It too appeared in the *Yale Law Journal,* under the title "Legal Planning of Petroleum Production: Two Years of Proration."[3]

As we re-read these articles in the years that followed, we both felt we had little to be ashamed of. Although we pretty much described how the industry operated and hit the mark on what could be done about some of the evils we saw, we did make one horrendous mistake. Since an honest confession is supposed to be good for the soul, I'll tell you about it. I still shudder when I remember one sentence we composed. To solve the problems in the oil business, we concluded, price-fixing must come to the industry. Years later price-fixing came, but it solved nothing. In fact, it only made matters worse.

In the same year our second article was published, and just months after we both joined the Department of the Interior, Secretary Ickes asked us to draft an order under the Code of Fair Competition for the Petroleum Industry (Oil Code) which would fix the price for every grade of crude oil and refined petroleum product at every point in the United States. This brought us down to earth with a jolt. We tried to do the job, but finally we had sense enough to tell the secretary we didn't know enough to devise a national price-fixing order—and no one else did either!

We persuaded Ickes that if we tried it, we probably would end up with monumental black markets and a legal nightmare which neither we nor anyone else could administer. Even though the drive in 1933 was to fix price floors rather than ceilings, the administrative nightmare that occurred after Congress and the Department of Energy set price ceilings in the 1970s seems clearly to demonstrate how right we were in saving our secretary from this fate.

But again we're jumping ahead of our story—let me backtrack for a moment. My association with Harold Ickes—the self-proclaimed old curmudgeon of New Deal fame—came after a telephone call following Franklin Roosevelt's taking office as president in 1933. In May of that year, the phone rang in my rather sumptuous assistant dean's office at Yale, and it was Ickes, the new secretary of the Interior. He said he'd read my articles about the petroleum industry and won-

---

3. J. Howard Marshall and Norman Meyers, "Legal Planning of Petroleum Production: Two Years of Proration," *Yale Law Journal* (Feb., 1933): 702–47.

A National Guardsman checking some well valves after the East Texas field was shut in by martial law. *Courtesy East Texas Oil Museum*

dered if I would join the Interior Department for the summer to assist with some of their oil problems. Since the offer seemed better than teaching summer school, I accepted, with the condition that the secretary agree to transfer my old colleague Norman Meyers from the Federal Power Commission to Interior to join me. Ickes obliged, and suddenly the whiz kids who had written about oil became assistant solicitors of the Interior Department and were thrown squarely in the middle of one of the most tumultuous times in the history of the oil industry.

Perhaps I had some premonition I would never get back to Yale. I remember wondering as I closed the office door at Yale whether I would ever again sit in as elegant an office. A hand-carved desk, oriental carpets on the floor, stained glass windows with lead panes—

all were part of the Sterling Memorial Law School building with its long hallways and Gothic architecture. The building was once aptly described by Dean Hutchins when he said: "I asked for a place to work and got a goddamn Gothic bowling alley." Ironically, the building was a gift of John W. Sterling, old John D. Rockefeller's attorney! I often wondered if he and his boss ever rolled over in their graves at what came out of his building, including Norman's and my articles on oil.

Only academicians from wealthy schools and high-flying bureaucratic chieftains can afford such luxury. Certainly I never came close to affording it until I became a bureaucrat—Chief Counsel and Assistant Deputy of the Petroleum Administration for War in the new Interior Building. There my big corner office and private washroom were even better than what I'd had at Yale, where I had to take two steps across the hall to reach the faculty restrooms. At least at Yale, however, none of my students had the temerity to post a sign over the commode, as did my government staff, which read "The Interior Department Suggestion Committee asks that you find ways to do it better and quicker."

CHAPTER TWO

Regulating Oil at Interior, 1933–35

When Norman Meyers and I were sworn in as assistant solicitors of Interior in June, 1933, neither of us had the foggiest idea of what we were in for. In Texas in those days, it seemed, everyone had a title. Lawyers of any prominence, regardless of any real judicial experience, were called Judge. Anyone who had ever been remotely connected with the military, including the local militia or the National Guard, was always called Major or Colonel. Doctors of any kind were a dime a dozen. Since Norman Meyers was an honest-to-God Ph.D., and I was at least a legitimate assistant dean on leave from Yale, these young gold-dust twins became known, at least in Texas, as "the Doctor and the Dean."

The first assignment the secretary gave me was to prepare a legal opinion that oil companies accounting for governmental royalties due were not entitled to deduct an arbitrary 3 percent B.S.&W. allowance. This opinion was easy to write and to defend since, although 3 percent might have been the right figure once upon a time, modern pipelines and centrifuge measurement had made the correct figure something more like 0.5 percent. There was one embarrassing moment, though, when the secretary archly asked me what B.S.&W. stood for. I answered that it was bottom sediment and water. "Really?"

The "36 Section" of the Elk Hills field in Taft, California.
Courtesy Chevron Corporate Library

he pressed. I answered that that was correct, "except, sir, in the oil fields—there it's called bull shit and water." I always believed he had put me on the spot on purpose.

The Section 36 Decision

The next task was not so easy, however. Hearings had been held recently on the famous "36 Case," and the evidence was in. By virtue of an 1853 federal statute, each state received from the government each section 16 and section 36 from the public domain as a contribution to its school funds, unless—and it was a big "unless"—such sections were known mineral lands at the time the public domain was surveyed. Certain lands which were part of the public domain in the San Joaquin Valley in California, for example, were surveyed in 1903. These lands included section 36, within which the Elk Hills field (Kern County) was located. No oil was found in this section (or anywhere else in Elk Hills) until about 1919. In fact, section 36 sold for a few dollars an acre for delinquent taxes some years after it was first thought to have been acquired by the State of California, which, at the time, valued it only as raw desert land. Standard of California then bought the section from third parties, and, around 1919, Standard drilled a "wildcat," bringing in the Elk Hills field. Question: was it "known" mineral land in 1903?

It was clear to Norman Meyers and me that it had not been known as mineral land in 1903, and we so advised the secretary. Now that time has run its course, the truth can be told. We had a solicitor of the Interior Department who looked at the 36 Case as a route to a federal judgeship. With the legerdemain for which lawyers are famous, he wrote an opinion shifting the meaning of the word "known" to "knowable." In other words, somebody *should have* known the land's potential. Elk Hills today would be recognized as an obvious anticline. But the anticlinal theory of oil accumulation had not even been invented by 1903. The secretary signed the opinion that the solicitor wrote, and Norman and I let it go by the boards. As I found out later, when a case is headed *The United States v. The Standard Oil Company*, if you are an attorney for the defense, you have almost struck out before you start swinging your bat. Alas, the poor solicitor never got appointed to anything better than a police court judgeship in the District of Columbia. But from it all came a wonderful jingle written by Ickes's old law partner, Donald Richberg. Playing on different pronunciations of Ickes, it read:

> On 36 the poppies grow
> Between the derricks row on row
> Black gold that once was Standard Oil
> Until the New Deal came to foil
> The sixty families and their tricks
> With law made up by Harold "Ikis"
> Tell the Marines, tell all the Mikes
> They must have known, says Harold "Eyks"
> Contrary proof can never stick us
> "We are the judge," says Harold "Ickus."

There came a strange sequel to the famous 36 Case. Under the 1853 statute, remember, it was not only each section 36 that went to the state of California, but also each section 16. The Elk Hills' section 16 was sold by the state and bought much later by the General Petroleum Corporation. Even though it was pockmarked with dry holes and has never produced a barrel of oil or a foot of gas, it too was judged to have been "known" mineral land in 1903. This ridiculous result happened because Al Weil of General Petroleum, who along with Oscar Sutro of Standard had argued the 36 Case, forced the government into a corner, and the idiocy of the first decision became even more obvious. To such extremes do well-intentioned administrative officials go—particularly if their opponents happen to be large corporations. After all, it is easy to forget that these same corporations are partly owned by widows, orphans, and the general public, whom administrators are sworn to serve.

Almost twenty years after having taken the risks of drilling and finding the great Elk Hills oil field, poor old Standard had to account for every barrel they ever produced from section 36. It cost them $7.1 million in cash and $4.5 million in developed assets and would have broken an independent. Had it been independent, on the other hand, the decision might have been different.

Regulating Oil under the NRA

Fortunately for Meyers and me, other, more pressing problems relieved us of the necessity of trying to rationalize an irrational section 36 decision. During the hot summer of 1933, the New Deal was seeking to find routes of recovery from the disastrous depression of the early 1930s. One of these routes was the National Industrial Recovery Act, sometimes known as the NRA and symbolized by its logo, the Blue Eagle. Several sections of this act had oil applications, and Section 9(c) applied specifically. The president, after a proper finding, was empowered to prohibit the movement of petroleum and its products in interstate commerce if it had been produced in violation of state conservation laws. Other sections dealing with the formulation of Codes of Fair Competition and regional trade agreements were soon to be applied to the petroleum industry.

Formulating Codes of Fair Competition for a multitude of different industries was delegated to the National Industrial Recovery Administration, headed by retired general Hugh "Iron Pants" Johnson. For weeks General Johnson presided over protracted hearings in Washington. Ickes assigned Norman and me to attend the hearings, giving us instructions to figure out how to take the administration of the oil code out from under the general and put it under the Secretary of the Interior. With the help of various oil men—both independents and majors (Oscar Sutro of Standard of California and Barney Majewski of Deep Rock Oil, in particular)—we managed to steal the general's oil show by persuading the president to transfer the administration of the oil code to Interior.

It really was not too difficult to accomplish. As usual, the industry was hopelessly divided. A few majors and most of the independents wanted price–fixing by the government—prices fixed *up*, not down. Most of the larger majors wanted none of it. We sided with the price–fixers, who probably thought they could get more of what they wanted from Ickes than from the general. Our side prevailed, and, by the end of the summer, everything related to oil that fell under the mandate of the NRA was transferred lock, stock, and barrel to the Secretary of the Interior—or, as it turned out, to his two ambitious underlings, who thought they knew exactly what to do. Later we could only boast that it did not take us long to learn that we did not have pat answers to complex problems.

All administrative agencies must have "Boards," and Interior promptly established the Petroleum Administrative Board with the solicitor of the department as its chair. Fortunately, he was too busy with the 36 Case and other matters to bother us much. Naturally Norman and I put ourselves on the Board, along with Charles Fahy as vice-chairman. Fahy, then a senior assistant solicitor of Interior and we two youngsters from Yale pretty much ran the show until the NRA was declared unconstitutional in May, 1935.

What kind of show was it? It was hell on wheels. The Code of Fair Competition for the Petroleum Industry was a mass of ambiguous, conflicting rules "voluntarily" adopted by "the industry" (whoever they were), which sought to set forth codes of conduct relating to a vast array of competitive minutiae, from the wellhead to the service station. While not nearly as bad as the stuff that emanated from the Department of Energy prior to decontrol in 1981, it was bad enough to drive each of us a little crazy. It never really worked and would have fallen of its own weight, if the Supreme Court had not mercifully ended its agonies in the Schechter Poultry ("sick chicken") Case— another instance where the court was right for the wrong reason.

In a lengthy opinion, the court held that the whole NRA was an unconstitutional attempt to regulate intrastate commerce. It conveniently overlooked the doctrine of "in or affecting interstate commerce." A year or two later, when the court upheld the National Labor Relations Act (clearly a regulation of local activities), it remembered the "in or affecting" rule and curtly dismissed, without ascribing reasons, the line of cases upon which the sick chicken had been based. I suspect it was the last time the Supreme Court ever concerned itself with whether any type of conduct was "in" or merely "affected" interstate commerce. Today the court regards everything as "affecting," so the federal government can regulate almost anything—and does!

This decision came down about the same time as the attempted court packing by FDR. It was later remarked that while the Supreme Court is supposed to act from a carefully insulated chamber, this 5-4 decision meant that at least five of the justices must have had their ears to the keyhole. Otherwise, how would they have known how they were threatened from the outside world?

Although the Code of Fair Competition was a monstrosity, I shall be ever grateful for its existence. It forced us bureaucrats to learn something of the competitive complexities of the business. What seemed so simple back at Yale turned out to be anything but simple. One of the first things we did to cope with these complexities was to establish a blue-ribbon industrial advisory group called the Planning and Coordinating (or P&C) Committee, sometimes referred to as the "Peak and Chisel Committee."

The idea for the group came from Jimmy Moffett, a former vice-president of Standard of New Jersey, who had fallen out with his company. He took the idea from the old Petroleum War Service Committee of World War I. (Later, Ralph Davies and I borrowed the same idea to establish the Petroleum Industry War Council during World War II.) The War Council did a great job, but the P&C never did—probably because, as always, the industry was so highly competitive and there was no war to force a competitive cease fire. As Davies once laughingly put it, when tankers from two different companies collided at night in the early stages of World War II: "Don't those fellows know there is a war on? They ought to lay off that competitive stuff."

Jimmy Moffett was one of the leaders of the price-fixing faction and promptly precipitated that issue in the P&C Committee. After many a wordy battle, a majority of the committee recommended to Secretary Ickes that he fix prices throughout the entire oil business.

Although many forgot about it later, fiercely competitive executives actually asked the government to fix prices—floors, not ceilings—crying "save us else we perish." Those of us in government attended the meetings and listened, and a more frightened lot of oilmen you never saw. The president of Barnsdale Oil Company carried his most recent balance sheet in his pocket, showing it to everyone and wearing an agonized expression on his face. Norman and I persuaded J. Elmer Thomas, an independent oilman and financial analyst, to attend P&C meetings with us. He sat across the room, and we told him to scratch his face if any of the factual arguments were off-base when the boys discussed "costs." After a week he'd almost scratched off his beard.

"Dollar oil" became the slogan at a time when the market dictated twenty-five cents a barrel for legal crude oil and ten cents for illegal hot oil in East Texas. The price-fixers were desperate and strong. They labored under the illusion that, despite the enormous surpluses of oil and its products, all that was necessary was to persuade the secretary of the Interior to wave a magic wand and have his lawyers draft an order fixing the prices for all crude oils and petroleum products at every point of sale in the United States. Then, all would be well. Of course they described it as prohibiting sales below "cost"—whatever that meant. Such a recommendation went forward from the Industry Committee to the secretary, supported by Moffett and Mike Benedum, an independent Pittsburgh oilman.

Secretary Ickes also believed that a nationwide price-fixing order should be issued, and he directed his lawyers to prepare one in September, 1933. Suddenly the authors of "Legal Planning of Petroleum Production," who had written in the *Yale Law Journal* that price-fixing must come to the industry, were confronted with the task of doing what they had said should be done. It was a rude awakening. We tried. We finally enlisted the help of our economist oil friend J. Elmer Thomas, and then we all three tried. At the end of some weeks we went to the secretary to confess we didn't know enough to write a price-fixing order which would have any chance of operating effectively. I added that I doubted if there was anyone in the country who knew enough to do that. The secretary insisted, but we finally prevailed.

We won with an old lawyer's trick. We argued that, since we could not defend the proposed order factually, we could not enforce it legally. Trying to do so would lead to the worst series of black markets ever seen. That settled the question, and we set about looking for other ways to get the government and the oil business out of the mess they were both in. Looking at a similar situation some four decades later, when the Department of Energy and its predecessor agencies tried to fix oil prices throughout the nation, I can only point out that they got themselves into a black market mess with unenforcible, ambiguous orders.

An amusing aftermath followed our persuading the secretary not to issue a general price-fixing order. It illustrates so perfectly how a political system often works. One day a letter came across my desk addressed to President Roosevelt from Mike Benedum. I looked at the routing stamps on the letter: "Referred by Louise Henry McHowe [one of the President's political aides] to the Secretary of the Interior for acknowledgement and consideration." Another stamp: "Referred by the Secretary of the Interior to the Chairman of the Petroleum Administration Board for acknowledgement and consideration." Still another stamp: "Referred by the Chairman of the Petroleum Administrative Board to the Vice-Chairman." Finally, a penciled note from the vice-chairman—"Howard, will you please answer this—[signed] Charlie." I read the letter addressed to the president. It said that there were two young whippersnappers in the Interior Department named Marshall and Meyers who were preventing the oil code from becoming a "beacon light" in the recovery program. Would the president please see that they were fired forthwith?

The government being what it is, I took a piece of White House stationery from my top drawer and dictated a letter for the president's signature. He thanked Mike for bringing this matter to his attention and hoped that he would always feel free to communicate directly on any matter that gave his friend any concern. The answer I prepared naturally promised nothing. I knew the answer would backtrack the same chain of command through which it had come to me, with each actor in that chain surnaming the first carbon copy. As it happened, I was sitting beside Mike Benedum in a meeting when a White House messenger delivered the letter which I had drafted and the president had signed. I did the unpardonable—I managed a furtive look at it. None of my words were changed. Some years later, when I attended Mike's ninetieth birthday party in Pittsburgh, I told him the story. By then we were both wiser and had a good laugh.

The business of getting government out of the mess took something more practical than price-fixing orders. This was as true in the case of surplus oil in the 1930s as it later proved with shortages in the 1970s. In the 1930s, some of us stumbled onto a simple fact. If we could diminish the surplus, prices would take care of themselves—perhaps not up to the magical dollar a barrel, but something better than ten or twenty-five cents a barrel. The key was to bring about reasonable rates of production in the great flush fields that had been discovered in the late 1920s and early 1930s—the big-three being Oklahoma City (1928), Kettleman Hills, California (1928), and East Texas (1930–31). All were running wild—blowing their heads off, as a matter of fact. These and other, similar fields were the real source of our problems—ridiculous, short-run feasts with the ever-present threat of famine to follow when enough of the gas coming with the oil had been blown into the air. The result was oil above the ground ruthlessly wasted through haphazard storage and far more left underground, probably never to be recovered at any cost. Now, these were prohibition days in Texas, and a bottle of bootleg beer cost thirty cents. A sign on an East Texas oil lease proclaimed, "Three barrels of the best crude oil in the world for a bottle of beer."

Further south, in the big Conroe field, a careless operator had drilled a well which collapsed, swallowing the drilling rig and the derricks and spilling thousands of barrels a day into the resulting crater. It was sometimes called an "act of God well," since the accident allowed its owners to escape proration regulation as it then existed. For months the crater drained oil from offsetting owners while gas was wasted, dissipating the natural pressure in the reservoir necessary for maximum recovery. I flew over this crater in the fall of 1933, when it still looked like an active volcano. It was killed some years later, when an offsetting operator paid a big nuisance value for the privilege of drilling a directional well that came close enough to the bottom of the hole under the crater to permit the pumping of mud and water into the producing formation, which shut off further flow. How much was wasted in the meantime will never be known for sure. As DeGolyer said, "The industry has not much experience producing oil wells through a one-acre choke valve."

That airplane trip over the crater caused me trouble. Some "friend" of mine in Interior reported to Secretary Ickes that I had made the flight in a single-engine puddle–jumper belonging to the Humble Oil Company (then a subsidiary of Standard of New Jersey), and I was called on the carpet. I told the secretary it was true, but that I had made the trip in keeping with my policy of associating with everyone in the business—big, little, and intermediate. My "friend" should also have reported that on my last trip from Tyler to Dallas to get back to Washington, I had borrowed the plane of Clint Murchison, a prominent independent in the area. In those days the secretary even got an order that said, in effect, "Thou shalt not eat with the industry." I resisted, since I learned so much over luncheon and dinner tables. Norman Meyers and I told this great boss of ours that if he thought we could be bribed by a free lunch, he had the wrong men in the job. He agreed, obviously because he trusted us. One of his nicknames was "Honest Harold," and he was proud of it. Certainly no money could buy him, although I once told him I was not sure that was true of a newspaper headline. But, after all, he was a politician—his sharp tongue and vitriolic style made him one of the most feared adversaries in Washington. A master of the informed leak, he was really tough only on paper or behind a microphone. As a boss he was really great.

Even before the whiz kids from Yale landed in the middle of the oil field melee, certain oil states were attempting to cope with the irrational and chaotic conditions in the fields. Oklahoma, Texas, Louisiana, Kansas, and New Mexico passed so-called proration laws by 1935 to limit production to "market demand." Lesser oil states, such as Arkansas, Mississippi, and Michigan, soon followed suit. The stated purpose of such laws was twofold—to prevent both above- and below-ground waste and to protect the correlative rights of competing owners drawing oil from a common pool. Unless some equitable basis was devised to divide a field's production among all the individual operators in the field, each would seek under the law of capture to take whatever possible as quickly as possible and let the devil take the hindmost.

Long were the debates as to what was "equity" in oil production. Should the division be based upon a percentage of the open-flow potential of each well? Or upon the relative number of surface acres which each operator owned in a particular field? Or some combination of acreage, oil-sand thickness, open flow, gas-oil ratios, "physical" waste, "economic" waste, maximum efficient use of reservoir energy, or something else? Lawyers and engineers fought before state commissions and in the courts, all driven by the self-interest of their clients.

Bob Hardwicke of Texas, a lawyer who also understood petroleum engineering, said that proration had led to a new professional—a "lawgineer"—which combined the worst characteristics of both professions. Each new formula for dividing up the pot was promptly weighed—not so much on its merits as on its effects on the interests of a particular producer or class of producers. The bottom line dictated each company's "philosophy."

No matter what allocation formula was attempted at the state level, nothing seemed to work. The respective governors invoked martial law for first the Oklahoma City and later the East Texas fields. They ordered in the National Guard to enforce proration, but, as time went on, even the troops failed to stop the illegal flow of crude output. The courts declared the proclamation of martial law illegal, improper, and unconstitutional.

The problem transcended state lines. If one state regulatory body did a good job of conserving, it might lose out to another state more interested in a quick buck. Years later, the process was repeated on a global scale between nations in the Middle East, Africa, South America, and the Far East. Their battles led to the creation of the Organization of Petroleum Exporting Countries (OPEC); ours led to an Interstate Compact to Conserve Oil and Gas and the Connally Hot Oil Act. Today, OPEC hangs together because Saudi Arabia, controlling the biggest reserves, can make it or break it. So long as the Saudis are willing to cut back output to make up for the overproduction of their neighbors, OPEC will probably survive. In the 1930s, so long as Texas was willing to take it on the chin, and imports were limited by tariffs, market-demand proration limped along.

Throughout the midcontinent of the United States in the early 1930s, lawyers, legislators, economists, engineers, and academicians argued about whether proration to market demand was really merely a scheme to raise prices, or, in fact, a necessary basis for achieving conservation by preventing physical waste. Oil operators could not afford to produce fields correctly at ten or twenty cents a barrel without going broke. On the other hand, if proration could mitigate the rigors of the law of capture, prices would tend to rise in the short term and stabilize in the longer term. These simple, yet interrelated economic and physical facts—unique to the producing end of the oil business—do not seem to be understood even today by classical economists like M. A. Adelman of M.I.T. or Milton Friedman, long associated with the University of Chicago. Unlike my old economic mentor, Walton Hamilton, who believed that no universal laws of economics existed, these gentlemen seem to think in generalities rather than in the specifics of a particular business. To express the controversy in simple terms, the classical economist seems to have missed the point that an oil field is somewhat like a bottle of soda water. Remove the cap slowly and hold back the pressure with your finger, and you can squirt a lot more of the liquid out of the bottle than if you jerk the cap off and allow the gas to come forth with a rush, bringing with it a lot of foam but not too much liquid.

All of the foregoing brought Meyers and me back to the question of what could be done to bring some order out of chaos. We had convinced our secretary that neither price-fixing orders nor marketing codes would do the trick. How about trying to make proration really work back in the oil fields? During the early summer of 1933, we made a first futile attempt. Seizing upon Section 9(c) of the NRA, we created a set of regulations directed at all transporters of oil and oil products in interstate commerce. They were so tight that hardly a barrel of anything moved for a few days, until we backed off the utterly impractical sections of the regulations. We must have looked as bad or worse than the Department of Energy regulators looked in the 1970s (with so many of their rules and regulations), and we failed for precisely the same reasons they did. We didn't understand the ramifications of our actions for the industry we were seeking to regulate. At least we were on the right track, however, and, in the fall of 1933 and the early months of 1934, we came back to the notion of controlling supply by making proration work—at least to a certain extent. At this point East Texas and California were the hot spots. Texas had a proration law that too few obeyed; California had nothing except the law of the jungle.

Fighting Hot Oil in East Texas

With the ability to satisfy one-third of national crude-oil demand at full throttle, East Texas drove the national market. We had to tame this field or go home.

We set up an office in East Texas. We wrote new and (we hoped) better regulations requiring producers to submit affidavits with each shipment of crude or oil products, swearing that the oil had been

A common East Texas practice—a bypass valve diverting oil from a legitimate to an illegitimate operator. *Courtesy East Texas Oil Museum*

produced in accordance with the rules and regulations of the Texas Railroad Commission (the regulatory authority attempting to administer proration in Texas). The secretary assigned the Division of Investigations to help enforce our regulations, but the division failed. Affidavits were soon worth a dime a dozen, and often they were signed with fictitious names like "The Guess Who Oil Company" or "The Amicus Curia Refining Co." The story was told of an oil executive found signing papers without having first read them. When his sales manager told him they were sales contracts, he exclaimed, "My God, I thought they were affidavits."

I acquired another title to join assistant solicitor—special agent of the Division of Investigations. About the only good it ever did was to get me out of a speeding ticket, when I was picked up by a state trooper outside Austin doing about eighty miles an hour. I handed the officer my special agent card and told him I was on official business. He saluted me and apologized. I somehow kept a straight face and said what I had never said before but always wished I could—"That's all right officer, but don't let it happen again."

The East Texas hot oil situation provided me my first opportunity to meet a genuine President of the United States, when I accompanied Secretary Ickes to a meeting with President Roosevelt. FDR asked about the situation in East Texas, and the secretary told him that I was about to leave for that area and would report shortly. I was not back in the president's office until some months later, and, as I passed his desk, he looked up and said, "Hello, Mr. Marshall. How did you find things in East Texas?" What a master politician! For him to call me by name and remember our previous conversation seemed nothing short of miraculous. I later learned how he did it—he had a staff to remind him—but that he considered it important marked him as rather extraordinary, even for a politician.

Much has been written about the horrors of hot oil in East Texas. (Hot oil was a nickname for contraband oil produced in excess of state allowables.) If I started to relate all my experiences with it, they alone would take a separate book. One, however, stands out in my mind. The only time I've been shot at in my life came late one night in an East Texas hot oil refinery. I was with Capt. E. N. Stanley, the chief enforcement officer in the field for the Texas Railroad Commission. He was an old regular army officer and a friend of Col. E. O. Thompson, the new chairman of the Railroad Commission. After someone started shooting at us, Stanley drew a big six-shooter and started shooting back. He went on into the refinery, while I hid behind the biggest tree I could find. He had been restrained by a state judge from checking a big fifty-five-thousand barrel tank suspected of containing hot oil, and I had proposed we deputize him as a federal officer to get around the state injunction.

With literally thousands of wells in the East Texas field hidden in the piney woods of an area fifty miles long and eight miles wide, it would have taken more than the whole Texas National Guard to police the area. Most of the wells could produce thousands of barrels a day. Theoretically, they were restricted to a few hundred barrels a day, and an intricate system of secret pipelines, valves, bypasses, and bogus wells, as well as crooked gangsters straight from Chicago, posed what seemed to be insolvable problems. Still worse, about a hundred very small "tea–kettle" refineries sprang up all over the field to distill the illegal crude. They piped, trucked, stole, and otherwise acquired illegal oil from one end of the field to the other. Once the oil had been refined in a tea kettle, no one could be sure where it came from. They merely made a rough fractionation of the oil (a beautiful 38–40 gravity sweet crude) and shipped out crude tops and topped crude by tank car all over the country and around the world. The topped crude was useful oil for steam boilers and railroad locomotives. If refiners, or even producers of virgin crude, had no immediate market for the crude, they dumped the stuff in earthen pits. The light ends evaporated, and the heavy stuff was sold for a song.

If ever I saw a clear-cut case of physical, above-ground waste, this was it. The crude tops were just poor-grade gasoline and heating oils, and they went directly to market for a few cents a gallon or to better refineries to be finished into specification material. As refineries, these tea kettles were not much different from (and just as wasteful as) the little skimming plants that sprang up like mushrooms between 1974 and 1981 to take advantage of the entitlements subsidy the Department of Energy provided for so-called small independent plants. These subsidized plants, which also could not make high-grade products, wasted high-quality crude oils and were a monument to foolish government regulations.

The hot oil refineries of East Texas were the bane of our existence. One of them on the Gulf Coast, called Eastern States Refining, was a pretty good refinery. But we used to say that if it ever ran a barrel of legal oil, it would blow up. Many years later, when I was executive vice-president of Signal Oil & Gas, Eastern States was bought by my company while I was out of the country on another assignment. Its problems were dumped in my lap by my chairman for no better reason than that I was the only one in our company who had a long background of refining experience. The first time I saw the seller, Dick Kahle, we both laughed, for we'd crossed swords before. We recalled some of our battles. Long ago he had defined for me the different categories of hot oil. First, there was the hot oil produced in violation of the orders of the Railroad Commission. Then there was the hot oil produced in violation of the Commission orders, for which nobody paid the royalty owner. Then there was the hot oil where nobody paid anything to anyone. Although Dick, now the former owner of Eastern States, claimed he'd never run any of the last, I still wonder sometimes about the truth of that.

The prices for East Texas oil got very low at times. I once checked an invoice from a hot oil refiner showing the buyer had paid $25,000 for a million barrels of East Texas crude, which figures out to 2.5 cents per barrel. The "going" price for the hot stuff was 10 cents, compared with two bits for the "legal" oil.

Most of the major oil companies at least gave lip service to the rules of the Commission, though some participated in the racket as well. Many shut their eyes when it came to purchasing illegal material. Millions of clandestine barrels, for example, ran through the Atlas Pipeline to Shreveport, Louisiana, then through the Texas Company's line to Port Arthur on the Gulf of Mexico. This was so flagrant that it was said that the Texas Company line would soon have to be covered with asbestos to protect it from all the hot oil flowing through it.

One favorite device for running and selling hot-oil was the reclamation plant. Millions of barrels of so-called "tank bottoms," along with fresh oil deliberately dumped into creeks to be collected downstream, found their way to these dinky little plants that separated prime-quality crude oil from the water and sediment contaminating it. Even when the State of Texas appointed receivers to take over and operate the properties of convicted hot oil runners, this failed to stop the racket. Usually the receiver ran more hot oil than the hot oil operator.

Another circumvention was with "storage." The field was dotted with lakes of crude oil poured into earthen pits, from which the illegal crude accumulations were sold and fines paid. Yet the level of oil in the pits never declined; they were refilled at night from unknown sources.

A local head of the American Legion was burned to death when his automobile exhaust ignited raw gas, which had accumulated at a low point in the road. Nearby wells, including his own, were producing oil without a flare to dispose of the associated gas. There was no flare because the flame would have advertised to the authorities that a well was improperly producing.

One operator built a concrete pillbox over the control valves of a well and attached it to his residence. He refused to allow access to state inspectors, citing the axiom that "a man's house is his castle"; his castle was called "the fortress of Gladewater." Another operator rigged up control valves in his bathroom, where he would send his wife to turn off the well whenever officials arrived to check things out. Pipelines were erected on top of pipelines to steal oil, and one dry hole produced an allowable for five years by diverting oil from a nearby producer.

Incidents like the foregoing would be amusing if the results had not been so tragic. They represented competition gone mad—where the prices of oil became determined by the prices paid for the hottest of the illegal oil—the incremental barrel in the jargon of economics. It would not have been so serious had East Texas been a relatively small field. But it was the greatest field ever discovered in the United States, with billions of barrels pressured at a shallow depth. Wells could be drilled and completed in less than a week. These wells, if opened wide, flowed like artesian water wells at rates up to twenty thousand barrels a day. Some years later, when they had a celebration marking the discovery of the field, Henry Dawes, then president of the Pure Oil Company, listened to the speeches describing the field as the biggest and best ever discovered. Nearby, Pure had found the Van Pool—perhaps a half-billion barrels—shortly before East Texas was brought in. Henry probably thought he had the world by the tail, but the crash in the business resulting from East Texas almost broke Pure. When it came Henry's turn to speak, all he could say was that he felt like the hen in the barnyard who laid the biggest egg in the world. When the other hens congratulated her on laying such a monstrous egg, all she could say was, "It might be the biggest and best to all you, but it was a pain in the behind to me."

It was this East Texas tiger which we somehow had to grab by the tail if we were to have any chance of making an oil code work. Two special agents of Interior, Archie Ryan and Joe Leach, and I moved against numerous producers and hot oil refineries. We were quickly hit by a series of temporary restraining orders, granted *ex parte* (without hearing our side), by the federal district judge for the area, Randolph Bryant. The actions were brought by an old "country lawyer" named F. W. "Big Fish" Fischer. Big Fish made a fortune by taking his fees in well interests in return for representing more hot oil operators than even he could remember.

Big Fish and the judge were good drinking friends. I was taught at Yale that it was unethical to talk to a judge about a case before his court without opposing counsel being present. But faced with the

W. F. "Big Fish" Fischer, the country lawyer of hot oil fame. *Courtesy Random House*

realities of law east of the Pecos, I figured I had better have a few drinks with the judge myself and did so. I finally made two deals with him. First, he agreed he would not grant Big Fish any more temporary restraining orders *ex parte* without giving me a chance to argue. Second, he agreed to sustain a demurer to an indictment against a hot oil operator or refiner on the grounds of the unconstitutionality of Section 9 (c) of the Recovery Act, which would allow a direct appeal to the Supreme Court. My argument was simple: if the act was valid, we ought to enforce it against everybody; if it was not, we could scrap it. Without going into all the inevitable legal maneuvering, the issues were finally joined in the case of *Panama Refining Co. v. Ryan et al.* in which I just missed getting my name in Supreme Court history—I was the first name in the "et al." list.

Charlie Fahy, much my senior as an assistant solicitor at Interior and an experienced lawyer, came to Texas to argue the *Panama* case before Judge Bryant. After he argued preliminary matters, the judge closed the first morning by chasing him with questions on constitutionality. Without any advance warning to me, Charlie responded by saying that he thought he would have me argue these matters after the noon recess. Charlie didn't know that I had never argued a case in court in my life. I have never since spent such a nervous lunch hour. I felt the way I had when I walked out on the field for the intercollegiate championships as the goal tender for Haverford College. I felt that the whole fate of the NRA rested on my shoulders. But as in the old soccer contest, once I "got my hands on the ball" it went all right. Judge Bryant refused the injunction, and somehow I managed not to get cited for contempt for having the temerity to drag a few law books into that Texas courtroom to answer Big Fish, who seldom bothered to use law books.

I shall never forget one other frightening incident from those tumultuous days. It was chronicled in *Oil & Gas Journal* for February 22, 1934. Big Fish sought one of his *ex parte* orders from Bryant, and the judge set it down for hearing the next day. The fog was so bad in Washington that Fish probably thought none of us would make the hearing. But the secretary of the Interior got me a two-seated, open-cockpit army observation plane, and they dolled me up in flight gear, including a parachute.

We took off for Tyler, Texas, by way of Atlanta and Shreveport. We broke out into clear weather at Atlanta, refueled and headed for

Barksdale Field near Shreveport sometime after midnight. This was long before modern navigational aids, and the pilot had to find his way by using road maps. When daylight came, both he and I had to figure out where we were, but finally the road maps and railroad tracks got us to Tyler without incident. In the morning I was almost deaf from the plane's engines, but the case was argued, and Fish was checkmated once again. The next day when we took off for Washington, half the town of Tyler seemed to be assembled at its little dirt-field airport. The army lieutenant wanted to put on a show, so he climbed a few thousand feet, dived almost straight down, buzzed the field, and did a loop. I was almost as scared in that plane as I had been in the courtroom.

But if the fires of illegal oil and flaming flares were hot in East Texas, the political fires in Washington seemed hotter still. Initially, we'd been able to check hot oil in East Texas because the bad guys labored under the illusion that the "Feds" had more power than we actually possessed. In Washington, no such illusion existed. The big companies, the little companies, the independents, the majors, the marketers, the refiners, the transporters, the integrated and nonintegrated companies, and the service station operators were, as always, hopelessly divided over what kind of power they thought we should have, and the old Petroleum Administrative Board in the Interior Department found itself caught squarely in the middle. Every one of the many warring factions wanted us to try to outlaw particular competitive practices which were thought to be "unfair." The oil code, theoretically, was adopted "voluntarily"; actually, it was adopted only after a series of compromises promulgated by government edict.

California without Proration

Having beaten back the price–fixers, what next? We hoped that some end to the wasteful and outrageous competition fired by the law of capture in the oil fields would bring supply and demand more nearly into balance, and that might make unnecessary the futile effort to regulate all the minutiae of marketing practices. California, however, posed a special problem. The other important oil-producing states—Texas, Oklahoma, Louisiana, New Mexico, Kansas, and Michigan—all had some sort of conservation laws. California did not, however. What California did or did not do affected other states, for not only did units within the industry indulge in cutthroat competition for a bigger share of a limited market, but the states also vied with each other. Even though the West Coast thought of itself as different from the rest of the nation, the same inflexible economic laws prevailed in the West as in the East, and what one region did had consequences for the other. We had to do something about California; the problem fell into my lap.

Beginning in the fall of 1933, I spent time in California. As in the Southwest, big fields had been found in California: Ventura Avenue (1917), Elk Hills and Santa Fe (1919), Huntington Beach (1920), Signal Hill (1921), Inglewood (1924), and Kettleman Hills (1928). I found no state laws or regulatory authorities to help us. The industry had tried previously to cope with the problems through a voluntary organization called the Central Committee of California Oil Producers. This committee, working through a so-called Oil Umpires Office, published what amounted to proration quotas for every well and field in the state. The quotas were as much honored by their breach as by their observance, but they were better than nothing—at least they were a place for us to start. If we could take some of the horse trading out of the quota assignments and construe a violation of the quotas as a violation of the Code of Fair Competition for the Petroleum Industry, then we had a chance to bring some order to the chaos of California.

To illustrate the kind of horse trading that went on, I'll recount the first meeting I attended of the Central Committee. It reminded me of a story told by J. Elmer Thomas, who had gone through "voluntary" proration in Oklahoma at Seminole. He said that whenever the engineers devised a new proration formula which was "fair and equitable," it was no sooner explained than all the oil executives reached for pencil to determine the effect on their own companies. So, at my first meeting of the California voluntary committee, the assistant oil umpire was explaining a new formula. No sooner was the formula on the blackboard than one of the committee members seized a pencil, ran some figures, and turned to me, a perfect stranger, with the words, "That can't be right; we lose four thousand barrels a day at Ventura Avenue."

I was told that all the operators at Sante Fe Springs previously had agreed as gentlemen to limit their daily production per well to a certain figure. A few months later, they had discovered that one of their number, George Machris of the Wilshire Oil Company, had failed to abide by his word as a gentleman. When confronted, he had answered simply, "You fellows know I'm no gentleman."

In an effort to devise some sort of formula that could be defended on an objective basis, Bob Allen, the assistant oil umpire and an able petroleum engineer, came up with the concept of the MER, the "maximum efficient rate" of production. The idea was based on the fact that if a producer attempts to extract oil from the ground too quickly in the short run, much less oil will be produced in the long run—a fact which some classical economists then and now either ignore or choose to remain "ignorant" about. We worked it out (more or less) and applied the MER on a "power curve," a mathematical theory which states that the greater the daily capacity of a well to produce, the greater the percentage of curtailment. This held back the big wells more than the small ones in order to make the system practical. Strangely enough, the new system began to work.

Still, we had some trouble with individual companies. George Machris's company not only produced and bought oil at Huntington Beach in excess of quotas set according to maximum efficient rates, but also, in effect, stole oil from the state tidelands by drilling crooked holes from the shore that bottomed under the Pacific Ocean, where they held not even any color of right. Another of these crooked hole producers, a dentist by trade, complained to my boss Secretary Ickes that I was persecuting him. The secretary asked me to see the gentleman about it. I got a simple explanation. He said of course his well had bottomed out under the ocean, where he had no lease but that the Supreme Court of California had ruled that oil belongs to whoever got it. I suggested if that were true, it might be easier to drill into somebody else's storage tanks.

Superior Oil Company, Union Oil of California, and Amerada Petroleum were classic cases where the self-interest of two or three individual operators producing wide-open from a few wells on a few acres under the law of capture can ruin a great oil field spread over thousands of acres. Superior alone, and Union and Amerada jointly, owned one quarter-section (160 acres) in one of the juiciest locations in the Kettleman Hills field, halfway between the gas cap on top and the oil-water contact on the flank. Save for these two quarter-sections, the whole 15,000 acres of Kettleman Hills had been unitized with the approval of the U.S. Geological Survey to provide proper well spacing and efficient rates of production. On each of these two lone quarter-sections, the companies in question had quickly drilled eight wells on twenty-acre spacing, flowed the wells wide open, blown into the air billions of cubic feet of gas, and sucked the water up from the edge of the field and the gas down from the top to create a pressure sink, from which the field as a whole never recovered. None of the three companies was an amateur in the business. All three knew the consequences of their conduct, but such was the nature of competition in this business. Each was quite willing to sacrifice the field and waste the resources so long as they got the "quick fix" of immediate production. How much oil and gas were lost or left unrecoverable in this great oil field we'll never know for sure. Unquestionably, the losses ran into the hundreds of billions of feet of gas and hundreds of millions of barrels of liquids—all "lost" by companies that knew better.

When the Supreme Court declared the NRA unconstitutional in 1935, I happened to be writing an order for Secretary Ickes to proceed against these three companies to stop the waste at Kettleman Hills. When I was told of the court's decision, I remember folding my file and remarking that I wished I could get away from all such tough problems so easily. Actually, I hadn't really gotten away. A few years later, I served as a private lawyer for Standard of California, and we were all still wrestling with the problem of Kettleman Hills. My senior partner, Felix Smith, and I were asked by M. E. "Tex" Lombardi, vice–president of production for the company, to try to come up with some legal method of stopping the waste that continued at Kettleman Hills. Felix and I both knew, however, that, under the law of capture and with the complete absence of conservation statutes in California to check waste, we were powerless to help.

Ralph Davies once told me that with all the twists and turns in this crazy oil business, someday you always meet yourself coming from the other direction, and so it happened to me with Kettleman Hills. The problem was still there when I was back with the government during World War II. Nobody ever really got rid of the Kettleman Hills problem until the offsetting operators at Kettleman paid Superior thirty-two million dollars in nuisance value to buy out their 160-acre lease. Even that came much too late to conserve anything like what had been wasted.

One or two good anecdotes about the Central Committee of

California Oil Producers and its oil umpire, J. R. (Bill) Pemberton, should be preserved for posterity. Bill was a brilliant, rough and tough oil geologist. He always defined an oil geologist as one who could tell you why you drilled a dry hole after you drilled it where he'd told you to. But Bill was better known in and outside the oil patch as a character. He had an incurable attraction for women, which more than once got him in trouble on the home front. When he took the oil umpire's position, a thankless job of trying to get his producer brethren to curtail output for the common good, he said it was to help him forget his domestic troubles. Some weeks before, he apparently had used up his last straw. When pressed for details, poor Bill could only mutter that his wife "walked right out on me and never gave me a chance to explain." We could guess what happened.

My late wife Bettye Bohanon, who used to work for the umpire's office and the Central Committee, had a story about Bill. After dinner one night, he asked her to spend a weekend on his big sailboat. When she said she could not, he asked why. "Because," Bettye answered, "I can't swim."

My best story came when I was trying to coach Bill as my witness in connection with an antitrust proceeding. I asked him how he would answer if the judge inquired whether he had had lunch on a certain day with Press St. Clair, the president of Union Oil of California. Bill said he would answer in the affirmative. Suppose he then inquired what they had talked about? Bill would reply, "Oh, the usual things, your honor, whiskey and women." He got the question and answered it just this way. The courtroom exploded. The opposing attorney was so flabbergasted that he dropped the subject. It was probably just as well the court was spared the details.

As for Bill Pemberton's luncheon companion, he was a tough old-time oilman himself. Once, after I had joined the California oil fraternity as a full-fledged member, I heard St. Clair object to a particularly funny and bawdy annual production of the California "Wild Cats" poking fun at the oil business and its quaint characters—a production written, by the way, by one of his own employees. St. Clair complained that it might undermine his "social standing." In fact, there were no halos in this bunch.

Somehow I survived my introduction to oil in California, primarily because I had sense enough to listen. As Sid Richardson, a legendary Texas wildcatter, once remarked, "When you're talkin', you ain't learnin' anything." Once, I failed to follow that rule. When the head of the Barnsdale Oil Company in California politely told me that, although his president in the East had told him that I was not available, he hoped that when I got ready to leave the government, I would talk to Barnsdale about a position with the company. "For what position—as a receiver"? I arrogantly wisecracked. It took me years to earn the respect of Barnsdale after this silly wisecrack. My remark had come too close to the truth for comfort, since they were skirting bankruptcy and receivership.

On the West Coast, as across the nation, an overwhelming surplus of crude oil completely disrupted the product market—particularly that money crop called gasoline. At that time the Los Angeles Basin represented the largest concentrated gasoline market in the world. Without much in the way of public transportation, in a city with millions of people spread over hundreds of square miles, automobiles provided the basic means of transportation. Some wag at that time posted a sign on the edge of Anarctica that read "Los Angeles City Limits."

"Price wars" flared with regularity. Tea-kettle refineries sprang up overnight, while some major refineries pretended that one new oil field did not exist because the crude contained too much sulfur. They "forgot" that such crude could be skimmed to make motor fuel, even if its high sulphur content gave it the name "skunk juice" and gave such a sour odor to the residual fuel oil that it would almost suffocate the passengers of the Southern Pacific when its locomotives passed through long tunnels. Today, three–tenths of one percent of sulphur is deemed "low." Low-sulphur fuel oil in the West in the 1930s was three percent, and much of what was burned ran as high as five percent.

Big tank trucks and trailer units compounded the marketing problems of the Los Angeles Basin. These big motor carriers, each holding as much as a railroad tank car, brought the spread of every oil well in the basin within ten cents a barrel of every refinery. Pipelines were cheaper though not by much, and they were not a necessity in California, as they were in the midcontinent.

"Big dump" service stations played on the markets like yo-yos. With hundreds of thousands of gallons of underground storage, big chains would sell gasoline a few cents per gallon under anyone else until their tanks were two-thirds empty. Then they would *really* slash prices—sometimes to the point where gasoline actually sold at the pump for less than the motor fuel tax per gallon. As a result, the whole market would fall, and all refiners, both big and little, then had to cut each other's throats. The "big dump" operators would refill their tanks for little or almost nothing and restore their postings to make a handsome killing until they were ready to repeat the process. As always, the buyer's market was attributable, a result of the frantic and wasteful way the crude oil was produced.

Someone once defined the best advertising sign as one that read "2 cents off." Nowhere in the world except in the Los Angeles Basin did we find great flush oil fields at the back door of a great metropolitan market. How could we deal with policies to regulate thousands of wells on the one hand and thousands of marketing outlets on the other?

We sought under the oil code to make each refiner-supplier of gasoline "responsible" for what happened to his gasoline in the marketplace. "Concubine" companies—marketing outlets that refiners "kept" but would not "admit" to having—were used to evade such responsibility. Shell used a company called Guardian, General Petroleum had Gilmore, Standard used Signal, and Tidewater-Associated owned the Seaside Oil Company. Others used jobbers, brokers, and every other kind of devious trick known in the business to avoid responsibility.

To illustrate the extent of the confusion in California, several anecdotes may serve. Under the oil code, it was more or less "legal" to agree about prices and "unfair" practices. Coming from a meeting of marketing executives, Charlie Jones, then the head of the medium-sized Rio Grande Oil Company, said to me: "Howard, this is a strange industry. There was not a man in that room whose personal check I would not take, but not one whose word I would take if he told me his price for gasoline tomorrow morning—including myself." At another such meeting, Al Weil of General Petroleum complained that Standard was dumping gasoline out the back door through its "concubine" company, Signal, to a cut-price operator selling under the name High Mileage. Late in the meeting, Standard produced invoices and truck tickets proving to all that the gasoline came from the Gilmore Oil Company, a General Petroleum "concubine." The next day a cartoon appeared at each man's office taken from *Esquire*. It showed a Dachshund wrapped around a tree sniffing his own rear. The Dachshund was labeled "Al Weil," the rear end was labeled "High Mileage," and the caption read, "Holy smokes, it's me!"

REGULATING OIL
Federal Initiatives and the Tender System

Through the long fall and winter of 1933–34, I seemed to spend most of my time on airplanes between Washington, D.C., Texas, and California. Many big shots in the business (who later denied it) prayed for us amateurs in the government to save them.

One of the later-forgotten prayers took the form of what became known as the Thomas-Disney bill, a proposal which would have invested the federal government with the power to regulate oil production within the boundaries of the producing states. The bill was sponsored as much by factions within the oil industry as by government. Some urged for it; others foresaw horrible consequences if the federal camel ever got its nose inside the oil industry tent, even on an emergency basis. In light of what happened years later under both Republican and Democratic administrations, perhaps their fears were justified.

To try to stop what appeared to be the steamroller called "federal control," the opposition launched a counterattack, the Interstate Oil Compact—a loosely defined confederation of oil–producing states. The organization aimed at keeping the federal government out of their hair. Most of the producing states subscribed, and Congress approved the coordinating agency in August, 1935. One notable abstention from the compact was California, which had not implemented market-demand proration.

The Interstate Oil Compact Commission provided a forum for the conservation and regulatory authorities of the oil states to meet and exchange ideas, make speeches, pass pious resolutions, and, in general, try to persuade each other to do the right thing—without much understanding as to what the right thing was. Over the years it has had about as much effect as most trade associations—nothing very bad, nothing very good. Because most so-called liberals and political trustbusters convince themselves that the oil industry is one grandiose monopoly, they have always argued that the Interstate Compact, even though it was a hodgepodge of state governments, was nothing but an instrument of monopoly. The cry resurfaced whenever congressional approval was asked to renew the compact (in 1937, 1939, 1941, and 1942, when the exemption was made permanent). Those who raise the cry have probably never read the compact or attended any of the group's meetings. A more futile window dressing instrument was never written.

The best that can be said for it was that it did stop the Thomas-Disney bill. As things turned out, this was probably just as well, although I thought differently at the time. Like most well-meaning bureaucratic regulators, we at Interior wanted power. We had little conception of the difficulties we would have faced trying to regulate all the oil fields on a national scale. We were lucky to lose this one.

Still, something had to be done. The solution to the problem was discovered, as usual, by accident. For me, the light dawned in a courtroom in Boise, Idaho, far from the oil fields. Eastern States, an oil refiner on the Gulf Coast which my company bought years later, had shipped two cargoes of gasoline made from illegal East Texas hot oil to the Fletcher Oil Company on the West Coast. Secretary Ickes ordered me to chase the two vessels (the *Republic* and the *Papoose*) with restraining orders to prevent their cargoes from being unloaded. My authority derived from the fact that the movement of the products of hot oil in interstate commerce violated Section 9(c) of the NRA. I chased them up the West Coast on a Coast Guard cutter, and I caught up with the *Papoose* late one night as it tried to sneak up Puget Sound to unload at Seattle. When the vessel refused to stop, the captain of the cutter threw grappling irons over the tanker's rail and cut his diesels, which emitted a streak of flames from the stack. The tanker captain, no doubt fearing an explosion, screamed that he was carrying gasoline and stopped. The Coast Guard officer and I climbed a Jacob's ladder and served the captain with a restraining order issued by a federal district court in Seattle. Not all was well, however, for the F.B.I. had let us down. They had failed to ascertain that the Fletcher Oil Company was not a corporation qualified to do business in the state of Washington but, rather, an Idaho partnership headquartered in Boise. Though I stalled in court for two or three days in Seattle, this jurisdiction problem got the case thrown out. But I hit the Fletcher boys with another restraining order in Boise the next day. By then, with both tankers having unloaded part of their cargoes in Seattle and Portland, Fletcher's storage in both places was contaminated with "hot gasoline." This shut down their terminals and left two tankers partly unloaded with demurrage running at four thousand dollars a day.

I shall never forget the trial on facts. I found myself alone at the counsel table and pitted against some of the leading members of the Seattle and Boise bars. Actually, it was to my advantage to be a lone government attorney without associates to have to consult with on each bit of strategy. On the first day I moved to cite the lawyers for contempt for having advised their clients to disregard the judge's restraining order. The judge agreed that they could not play fast and loose with his order, regardless of whether it might ultimately stick or be dissolved. This certainly got my opponents off to a bad start. Next, F. W. Fischer, the "Big Fish" of hot oil fame from East Texas, showed up in court. When I saw him, I shouted in a loud voice, hoping that the judge would overhear: "Well, well Big Fish—if I ever had any doubt about the heat of these cargoes, your presence in this courtroom is proof positive to me." The judge *did* overhear and later asked me in chambers what I had meant. I told him all about Big Fish—a highly improper maneuver, I fear, but a highly effective one. I'm afraid I had come a long way from my classes in legal ethics at Yale.

Then followed one of the most unusual legal proceedings to which I have ever been a party. My proof of the illegal character of the oil from which the gasoline had been refined consisted of a series of telegraphed affidavits from East Texas. My opponents, of course, objected to this kind of proof. I asked for a continuance so that I could travel to Texas to procure the originals, as well as other relevant material. The demurrage rate for Fletcher was four thousand dollars a day, so its lawyers naturally objected to my request. To "help" them, I offered to testify directly under oath how hot oil was run and to tell my personal knowledge of the business of the refiner who shipped the hot cargoes. Reluctantly the opposing attorneys agreed. So for two days I asked myself questions from the witness stand and subjected myself to cross-examination. I ended up with a favorable decision from the court. Rather than break the Fletcher Oil Company, however, I finally took a consent device which allowed Secretary Ickes to decide how the hot stuff would be gradually distributed.

Another strange incident occurred that I never could explain. Secretary Ickes called asking me to talk to Sen. William Borah of Idaho, who claimed I was persecuting good Idaho citizens. I called Senator Borah, and he claimed that the cargoes had been refined from legal oil. I asked the senator whether I had his word as a member of Congress and a lawyer that he knew this for a fact. Rather than giving it to me, he said he would call me back—he never did. When he died several years later and large sums of cash were found in his bank box, I could not help wondering.

Toward the end of the trial in Boise, I found myself wondering why we had to come hundreds of miles from the scene of the crime to prove that a particular batch of oil was hot. Why not reverse the order of proof and require the producers to prove it was "cold" before transporters were allowed to move any oil or products from the field? After all, there were only a few pipelines, railroads, and refiners, and they were unlikely to disobey our regulations. Here were the beginnings of wisdom: regulate by seizing the monsters at the throat—it worked.

We established what became known as the Federal Tender Board for the East Texas field in October, 1934. Anyone desiring to ship oil or oil products out of East Texas had to apply for a certificate of clearance—a "tender"—describing in detail what leases the oil came from.

We set up what amounted to a monthly book balance for each lease in the field. As fast as any lease exhausted its quota, that was it—no more tenders were available. Hot oil came to a thundering stop in a month. Alas, the initial victory was short-lived. When my old lawsuit from Judge Bryant's court in East Texas hit the Supreme Court on appeal, the court stripped us of authority to act. In *Panama Refining Company* v. *Ryan et al.*, the high court struck down Section 9(c) of the Recovery Act as an unconstitutional delegation of legislative power on January 7, 1935.

The case was argued by my old adversary, "Big Fish." He rented a coat and tails and put on the old "country lawyer" act. He even pulled a tattered copy of the oil code out of his back pocket, claiming he had searched far and wide just to find it. Of course the regulations under 9(c) were not part of the oil code, and neither Big Fish nor anyone else had any trouble finding them, but Fish got away with his act. The attorneys for the Department of Justice knew so little about our case that they failed to call his bluff. I sat in the courtroom, sick at heart, listening to allegedly able attorneys from the solicitor's office made into monkeys by a shyster from Tyler, Texas.

The Supreme Court spoke a few weeks later in an 8–1 decision, with Justice Benjamin Cardoza the lone dissenter. The majority opinion rested on the claim that the president had not made a proper "finding of fact" when he signed the order prohibiting the movement of hot oil in interstate commerce. While the case was being heard, one of my students from Yale, Bill Strook, was serving as Justice Cardoza's legal secretary. After the decision, he came over to see

Norman Meyers and me with a message from his boss. He said the justice had told him to tell us that next time we wrote an order for the president to sign, we should compile fifteen hundred pages of statistics, call them a "finding of fact," bind them in the blue ribbons of the department, attach a seal, and bring them over to a particular Supreme Court justice with a note saying, "You asked for it. Now, damn it, read it."

Fortunately, Fish's victory was too short-lived. A few days after the fateful decision, I had a call from Tom Connally, the senior senator from Texas, a real old-time spellbinder, string tie and all. In his office he asked me if I thought I could write a constitutional act to stop the hot oil racket. I told him I thought so—if I could have forty-eight hours. He seemed surprised. I went back to Interior, and Norman Meyers and I stayed up most of two nights writing what became known as the Connally Hot Oil Act of 1935. The senator introduced it, and it passed with only minor modifications. The Federal Tender Board was reinstated, and the ideas born in the desperation of a Boise courtroom finally became the law of the land.

Big Fish was finally bested. After a few more cat-and-mouse games early the next year, the flow of hot oil in appreciable amounts was finally ended once and for all. The Connally Act was tried, tested, and found to have firm constitutional foundation. Probably by this time, however, Big Fish did not much care. He had seventy–odd oil wells in the East Texas field, which he had taken in lieu of fees. He wanted no hot oil artists draining oil out from under *his* leases! He was many times a millionaire, and he even drove to his leases in a big Cadillac, which he named "Papoose." He said he paid for it with the fee the Fletcher Oil Company paid him for appearing a single day in that Boise courtroom on behalf of the tanker that inspired his Cadillac's name.

When we wrote the Connally Act, we tried to plug the leaks and loopholes that had let some hot oil escape. One such hole involved state proceedings that tied up great batches of hot oil which could eventually be sold off at auction by a receiver. Since only the hot oil runner had any idea whether he had good title to the oil (usually he did not care), the only bidder would be the hot oil runner himself for a few cents on the dollar. This became a slick way to "legalize" hot oil and make some quick money. In the Connally Act, we inserted an "in rem" proceeding, under which the federal government would confiscate hot oil. As in prohibition days there had been such cases as *The United Stated v. One Hundred Barrels of Whiskey*, we now had cases reading *The United States v. 100,000 Barrels of Crude Oil*.

Ten years later, when I was back again in Washington as chief counsel for the Petroleum Administration for War, a few hundred thousand barrels of hot oil were still sitting unclaimed in steel storage in the oil field. Some of those who probably produced it in the first place petitioned me to arrange for its release "to help the war effort." We did help the war effort—by arranging for the confiscation of the oil with the proceeds paid to the U.S. Treasury.

Before we leave the Connally Act, however, an interesting sidelight should be added, which illustrates so perfectly the old adage that there is nothing so permanent as a government bureau and its employees—"a few die and none resign." In March, 1935, the Federal Tender Board was reactivated within Interior to enforce the Connally Act. The bureau and its staff did just that and was reorganized as the Petroleum Conservation Commission in 1937; in 1942 it became the Oil and Gas Division, still within the Department of the Interior. Since the late 1930s, however, virtually no hot oil has been produced or run, but as of 1994 the bureau still existed. The enforcement staff has had literally nothing to do. I used to say that when somebody abolished the bureau I helped create, it would show that at least somebody in government really wanted to cut expenses.

With the creation of the Federal Tender Board in 1935, we were beginning to learn how to regulate—by applying regulations to the few at the passes rather than the multitude at the sources (or the even larger multitudes at the markets). We then pondered over California's unique oil situation. We had almost nothing to build upon except the voluntary efforts of the Central Committee of Oil Producers. We were saved by an idea from a young vice-president of Standard of California, Ralph Davies. There was a section of the NRA that exempted local trade agreements from the antitrust laws, if the exemptions were properly approved by the federal government. Under this section, the Pacific Coast Petroleum Agency and Refiners' Agreements were negotiated. Sometimes it was called the "California Cartel." Despite its wicked nickname, even the antitrust boys in the Department of Justice gave it their blessing.

The Cartel was simple in concept. In California, the narrow pass through which the river of oil flowed from wells to consumers was the refinery. The refiners were bound together under industry-wide agreements, pursuant to which the large refiners bought enough of the smaller refiners' surpluses to keep the minor companies in business. The stipulation was that these small refiners not provide outlets for the California equivalent of hot oil—that is, oil produced beyond the "voluntary" quotas established by the Central Committee, which were themselves subject to the approval of the federal authorities under the oil code.

The proviso worked. Why should a small refiner buy hot oil if not threatened by bankruptcy? Why sell gasoline to "big dump" service stations if you have a solvent buyer standing ready to pay a reasonable price? Of course, the system did not work perfectly. Some slippage, some chiseling, some foolishness was bound to occur. It did cultivate laziness among some of the smaller, tea-kettle refiners. Why work if you do not have to? Once I passed the nightclub in the Los Angeles Biltmore Hotel at high noon. A group of small refiners was enjoying a high old time rather early in the day. The director of the Agency Agreements, Bill Von Fleet, turned to me and said, "Well look at those independent refiners hard at work selling their gasoline." Neither Bill nor I was mad—just jealous. We had too much work to be out wasting our time like that.

Looking back, it is clear that the Agency and Refiners' Agreements did not eliminate but, rather, mitigated competition. All the refiners worthy of the name fought to hold their place and position in the market. Most producers of crude oil still battled to find new oil, develop leases faster than their neighbors, and argue for higher quotas in the Central Committee. Nothing much changed except the wild, woolly, and wasteful excesses of the law of capture and the monstrous marketing practices that followed from it. For a few short months in California, we had the wild beast partly tamed. Then the Supreme Court, in its ultimate wisdom, threw out the whole NRA in May, 1935. Afterwards, some of us in the private sector had the whole job to do over again. In doing it, we barely escaped with our legal eyelashes, but that tale comes in the next chapter.

With the West Coast in better (if not really decent) shape, the eyes of Interior turned once again eastward. Even though hot oil was being brought under control by the Tender Board and the Connally Act, an enormous backlog of gasoline, left over from the hot oil days, hung over the market like a cloud. Some in the industry thought that if coordinated purchasing of surpluses worked in the West, it ought to work in the East. So they set up a buying pool to sop up the gasoline surplus. The trouble was that they did it on their own, without formalizing a trade agreement and securing formal government approval as required by the NRA to assure themselves immunity from antitrust prosecution. Perhaps they were even encouraged by a naïve special assistant to Secretary Ickes to act without formal approval. Perhaps they feared that approval from the Department of Justice could not be secured. As much as Norman Meyers and I wanted the industry to do what had been done in the West, we specifically warned those setting up the buying pool that they were laying themselves wide open for future trouble, and our words proved to be prophetic. The trouble came in the so-called Madison Oil case two years later. William "Wild Bill" Donovan tried to defend against the antitrust charges brought against most of the oil industry east of the Rockies. He lost, though he appealed the decision all the way up to the Supreme Court. My old colleague on the Yale Law School faculty, William O. Douglas, wrote the opinion of the court. Legally, if not economically, he had grounds for upholding the charges. After all, the courts have never really been able to draw a very sharp line between "reasonable" and "unreasonable" restraint of trade. I will say more about the famous Madison case in chapter 3.

Leaving Interior

Thus far, with a few digressions, I have chronicled a series of events from 1933 to the middle of 1935, the heyday of the New Deal. It was now the late spring of 1935. The Supreme Court had thrown out the whole NRA in a case completely unrelated to what we were attempting to accomplish with the oil industry. One thing is for sure—the court wound up my first government career. I got the news at the same time I found out my wife was pregnant. As one of my friends said, "Isn't it funny how the unemployed always start to raise a family?"

So ended my first two years of government service—and so began the second part of my trilogy in oil. My last day at Interior was July 24, 1935, ending a great two years. Norman and I had had a chance to meet and know on a first-name basis almost every interesting oil executive in the nation. We got a practical education in a time of crisis, and that experience would be hard to duplicate. We had responsibility thrust upon us far beyond our years; we were both just in our twenties. Toward the end, I remember a phone call asking us whether we could spare time to have dinner with two elder statesmen in the business, Walter Teagle and Bill Farish of Standard of New Jersey. We accepted, of course. I turned to Norman to remark that this was fun while it lasted, but when our government jobs ended, we were going to have to go back to work in the real world.

The generation of oilmen that we had come to know well included Walter Teagle (of Standard of New Jersey), Harry Sinclair (of Sinclair Oil and Teapot Dome fame), Wirt Franklin (an independent who went broke after he discovered the great Oklahoma City oil field), Ken Kingsbury and Oscar Sutro (of Standard of California), Charlie Arnot (a long-haired piano player who headed Socony Vacuum), Press St. Clair (of Union of California), Al Weil (of General Petroleum), Bill Holiday (of Standard of Ohio, then an independent company with a major's name), Ed Suebert (of Standard of Indiana), J. Howard and Joseph Pew (of Sun Oil), and Otto Donnell and sons (of Ohio Oil). Another notable acquaintance was Henry Doherty of City Service, who not only headed his company but also wrote a classic treatise which recognized the virtues of one operating plan per reservoir—the plan was called "unit operation" or "unitization." Because he was the first articulate top executive who honestly recognized the need for efficient production practices in the field, he was regarded as controversial by his generation of executives.

A younger generation was coming up. They included Barney Majewski of Deep Rock Oil and Charlie Jones, later head of Richfield and Arco but then president of the independent Rio Grande Oil Company. Jones almost went broke because he found a big California field at Ellwood. Like Wirt Franklin, he had to obtain loans to drill up the field in order to hang on to his leases. Then the price of oil broke from a few dollars to a few quarters a barrel.

Then there was Bill Farish, one of the founders of Humble Oil, which Jersey thought it had acquired. In fact, Humble's first-generation executives ended up acquiring and revitalizing the management of Jersey. Thirty-year–old Ralph Davies, vice-president of Standard of California and recently returned from England, where he had been loaned to an affiliated company, was obviously a man with a future. His friend, Will Dodge from the Texas Company, was another. So too were Ed Pauley of Pauley Petroleum, Sydney Swensrud of Standard of

Ohio, and Paul Blazer, founder of Ashland. These and so many others taught us much that we hardly realized at the time.

Many in government also taught us. Senators, congressmen, attorneys with Interior and Justice and, most of all, that old martinet, Harold Ickes, secretary of Interior. The story was told of his first day in office when he surveyed his new domain to find an employee with his feet on his desk reading a newspaper at 9:30 in the morning. "Young man," said the secretary, "Is that the way to spend the government's time?" The young man looked up, and, not knowing who the secretary was, replied, "Well, what the hell is it to you?" He joined the five o'clock rush to leave the building, but never to return. Because live cigarette butts burned holes in the awnings if thrown out of the open windows of the old Interior Building, the secretary issued a general order subjecting any employee to instant dismissal if guilty of such an act. In those days I smoked and so did the secretary. Standing behind his desk one afternoon trying to explain the horrors of trying to fix prices, I forgot about the smoking order. I snapped my burning butt over his shoulder out the open window behind his desk. He promptly snapped his own through the same open window. The staff in front of both of us burst into laughter, and the secretary joined in.

The old boy, for all his puritanical exterior and sarcastic wit, had a great sense of humor. Once his staff presented him with a dossier on a proposed new assistant, explaining that not only was the man a hard worker, he did not smoke, drink, or go out with women. The secretary wrote across the dossier, "Is he also a virgin?"

Incidentally, the secretary himself had an eye for the women. During the war an attractive lobbyist almost sold him on building a concrete pipeline from Texas to the East Coast instead of using steel in what became the Big Inch lines. The deputy administrator and I always thought that, momentarily, the secretary's mind had wandered from the business at hand.

CHAPTER THREE

High Adventure with Standard of California, 1935–41

With the forced end of my government career and a pregnant wife came the question—what next? Secretary Ickes tried to persuade me to stay with Interior as an assistant secretary—a flattering suggestion for a mere youngster of thirty years. I pondered the offer, finally declined it and, when he asked me why, I told him that what he proposed seemed to me to be the start of a political career. I had learned that the further you advance in politics, the more vulnerable you become. Sooner or later, somebody asks what you have done for the party, and, unless you can command votes or a power base, out you go. On the other hand, if you embark upon a professional career, the further up the ladder you climb, the less likely you are to suffer someone pulling the ladder out from under you.

But if I were to return to law in the private sector, where and how would I do it? Out of the blue came a suggestion from Ken Kingsbury and Ralph Davies of Standard Oil of California (now Chevron). Standard at the time was a first-tier major, in a league with Jersey Standard, Indiana Standard, the Texas Corporation, and Gulf. Kingsbury, the president of Standard, who knew I was on a leave of absence from Yale, inquired whether I wanted to return to the school. I told him I thought I did, since I enjoyed teaching, but he asked if I might find it more interesting to join Standard. When I asked him what I would be doing, he replied, "We can use someone who knows both law and oil business. Oscar Sutro does but he is about the only one we've got. You seem to be a pretty good lawyer, and during the past two years with the government you have learned much about the oil business. If you join us, report to the eighteenth floor. Do anything they ask." I knew that the eighteenth floor was occupied by the managing directors of the company. It sounded fascinating—with great variety and many problems reaching into every aspect of the business. When I asked what the job might pay, he tossed out that we might "start at twenty-five thousand" (more like $250,000 in today's dollars). The government had paid me about a quarter of the figure Kingsbury had mentioned and Yale still less. Still, I told him I wanted to think about it.

I remembered the comment of Thurman Arnold, one of my colleagues of the law school faculty, about money. He described wealth as "earning as much or a little bit more than your friends"—not a bad definition. I talked to Charlie Clark, dean of the law school, who urged me to return to Yale, as he thought I would be his logical successor as dean. He would recommend a full professorship with tenure for me, and this laid in my lap one of the hardest decisions of my life. I finally resolved my quandary on the side of an oil career. As I explained it to the dean: "Charlie, if I come back to Yale, I'll be all right until some afternoon when I'll wander over to the Sterling Memorial Library to ask for the last six months of the *Oil & Gas Journal* to see what the boys are doing out there. Then I'll wish I were out there in it. Perhaps those who can, do, and those who can't, teach. If I come back to teach later, I'll probably be a better teacher after some years of tough experience." As it turned out, I would never return to the law school until I gave a guest lecture almost fifty years later.

Later in life, after a lot a experience, another oil-lawyer and I agreed that if we ever taught again, we would offer three courses, entitled: "Elementary Connivery," "Practical Connivery," and "Advanced Connivery." It would be hard to earn an "A" in this last course.

The offer from Standard reflected my work experience with both Davies and Kingsbury during my California trips on behalf of the Department of Interior. In retrospect, it was not surprising that they needed someone with my background to deal with rule-of-capture competition in the state.

STANDARD OF CALIFORNIA

I became an employee of one of the nation's largest oil companies on June 26, 1935. The original offer was just for the summer, but Kingsbury and Davies correctly figured that it would be much longer once I got into the flow.

As special counsel of Standard, I had a departmental manager's rating, which allowed me a private secretary, and I found one after sifting through a pile of personnel cards. He was working at a pipeline station at Estero Bay where he had some relatively minor job, and I noticed on his personnel card that he was a Phi Beta Kappa from Pomona College, which I knew to be a first-rate school. He had tried unsuccessfully to run a small business in the midst of the depression, then had moved on to learning shorthand and typing in a few weeks. I did not really need an expert legal stenographer; I wanted an administrative assistant who could think. After I talked to the man, I was sure I had found him. His name was H. Chandler Ide—the future chairman of Natomas and for many years the good right arm of Ralph Davies. He also coauthored the official history of petroleum war planning in World War II, *A History of the Petroleum Administration for War*. While I served directly on Standard's payroll, whenever I received notices instructing all managers to notify the members of their department of a holiday, I would call to Chandler in the next office, "Hey, department, we have a holiday!"

Chandler Ide had a knack with words, and a bit of verse about antitrust matters entitled "Mad Madison" still sticks in my mind. It ran in part:

> When price goes up it is a sin
> And critics raise a frightful din
> When price goes down there are no smiles
> For G-men start to search the files
> The world is in a sorry state
> When charts and graphs can seal men's fate
> And markets spot by their gyrations
> Can prove the crime of price fixations
> When oil men rot upon the rock
> And bankers buy up all their stock
> There will be no surplus then, I fear
> Of any worldly goods save beer.

Once I asked Chandler what a particular government order said. He replied, "The nouns give and the adjectives take away." Attorneys for the Department of Energy please take note!

When I left Interior, the secretary kept me on the payroll for a time, since I hadn't taken any holidays or vacations during my several years of service. Later it was said that this may have been the only time a man received checks for several months from both the government and an oil company legitimately.

State of Washington v. Standard of California

In the summer of 1935, "quo warranto" proceedings were initiated to throw most of the oil companies out of the state of Washington for alleged violations of its antitrust laws. The sin was not in raising prices but in cutting them. The term used was "predatory price cutting." It was alleged that the majors able to weather lower prices were doing so to put certain so-called independents out of business, the way the old Standard Oil Trust was supposed to have done. Regardless of what the old trust may or may not have done, our troubles arose from the same problems we had faced under the Code of Fair Competition. We had crude oil and refined products running out our ears. Competition was so intense, volume so critical, that even in a state hundreds of miles from the wild wells of California, gasoline was sometimes sold at the terminals for less than the state tax per gallon. Some even said we lost money on every gallon but made it up on the volume. No wonder it was difficult to explain. I was assigned the task of preparing a factual brief on the nature of competition in the oil business to be used by the attorneys for the defense.

Preparation of the brief took much of the summer. We planned to select a variety of expert witnesses to defend the different aspects of the brief. Before the case went to trial, Robert Hulbert, a dean of the Seattle bar and chief trial counsel for Standard in the State of Washington, took me aside. "You wrote this brief. With your academic and government background, I can qualify you as an expert. You take on the whole job, and if, after you have had a run at it, we think we need additional support we will try to get it."

The testimony started by explaining the competition for leases and geologic information in prospective and producing oil fields. It continued to town lot drilling; the way the law of capture forced wasteful, unrestrained production; the necessity of running refineries wide-open to take care of flush crude oil production; the pressure of overhead costs in a capital-intensive industry; and the requirement of constantly pushing refined products through pipelines, tankers, trucks, terminals, and service stations so that nothing interrupted the movement of crude oil from the wells to the markets, twenty-four hours a day, three hundred and sixty-five days a year. We got the trial judge so interested in each chapter of the story that he overruled most of the objections raised by the prosecution; he wanted to hear what was coming next. The story of the oil industry and the manner in which it is integrated, after all, is fascinating, complex, romantic, and absorbing, and I think the judge enjoyed hearing about it.

At the end of three days of direct examination, cross examination commenced. Mr. Hulbert again took me aside to say: "The Judge has let you wander all over the place. I do not intend to raise a single objection to any question the attorney for the State of Washington asks you. You are a lawyer; protect yourself." Probably with more arrogance than good sense, I told Mr. Hulbert that I would only feel I had done a good job for him if, whenever the state's attorney got the answer to a question, he would regret having asked it. After several days of cross examination, this was about the way it turned out. He made the mistake of cross examining a witness who knew much more about the subject than his interrogator, and when it was over, the judge leaned across the bench and announced he was ready to rule. All the oil attorneys rested their case in a hurry and breathed a sigh of relief when the judge threw the prosecution out of court.

That night at a victory dinner, Mr. Hulbert cut me down to size beautifully. He made a short speech, saying that, although he wanted to take nothing away from my performance, he wanted me to know how he would have handled me on cross examination. He would have asked me just two questions. "What did you say your name was?" The answer would have been, "Howard Marshall." "Who did you say you worked for?" The answer would have been, "The Standard Oil Company." Then, said Hulbert, he would have looked up at the judge, rolled his eyes to the ceiling, thrown up his hands and said, "That's all." Here was the difference between a real pro like Hulbert and a rank amateur like the prosecutor.

I learned other lessons, such as how to be an expert witness—which I always tried to remember when, over the years, I had to testify repeatedly before legislative committees or serve as an expert witness. Rule A, of course, is to know your subject and be prepared. Beyond that, never appear to be an advocate. Take it slow and easy, and pause before answering each question, even when you know the answer instantly. Take the time to try to figure where your questioner is going to go. Once, when I served as an expert witness for an interstate gas pipeline before an examiner of the Federal Power Commission, the general counsel for the line tried to coach me as his witness. Finally I told him, "I think I know your case and what you want me to try to do." "What is that?" he asked. "Argue your case from the witness stand." "Exactly," he responded, "but don't show it." That is what I did, but it did not appear so.

On the other side of the coin, Everette DeGolyer, a pretty good expert witness himself, told me of a case where the experts were trying to tell the lawyers how to try the lawsuit. In return, one of the lawyers told the experts he wished they would stick with their perjury and let the lawyers practice law—another way of saying that there are liars, damn liars, and expert witnesses.

For weeks in the courtroom in Olympia, we saw nothing but fog and rain. The clouds lifted on the afternoon of the last day of the trial, just before our favorable verdict came. It seemed symbolic that just outside the courtroom, Mt. Rainier in all its snowcapped glory appeared. The next morning, flying home to San Francisco, there was not a cloud in the sky. All the great volcanic peaks of the Northwest stood out in bold, snowcapped relief. What a way to end a lawsuit!

Gasoline Buying Pools

Winning in Washington did nothing to solve the problems that had led to the lawsuit in the first place. Was there anything that could be done legally in 1935 to end the shambles, which had almost immediately followed the decision of the Supreme Court invalidating the NRA and the codes and trade agreements pursuant to it? In the middle west, acting on generalized statements of the president to the effect that he hoped the gains of the NRA would not be lost, oil companies continued the gasoline buying program they already had in operation. A system of so-called dancing partners, so named by Charlie Arnot of Socony (speaking for the majors) and Barney Majewski (speaking for the independents), kept things on a fairly even keel until the Madison antitrust case blew it apart. With hot oil under real control for the first time, if each big refiner made it his business to "dance" with an independent refiner in the buying of gasoline (without any formal agreement between them), then the worst effects of cutthroat competition could be minimized. At least gasoline would not be sold for less than the tax, or (still worse) contracts would not allow the buyer to be *paid* for taking a tank car of gasoline to market.

In California, since a formal agreement (the Pacific Coast Petroleum Agency and Refiners Agreement) had been approved in the summer of 1934, it was deemed legally dangerous to play a "dancing partner" game after the demise of the NRA. A year or two before the NRA, California companies had tried a gasoline buying pool, when a broker named Frank Long bought gasoline from independent refiners on behalf of the majors in rough proportion to their total sales in the five western states. It was clearly illegal and, fortunately for the industry, fell apart. With the law of capture pushing relatively unlimited supplies of crude oil into the refineries, even the largest companies could not sop up the excess for long. No one wanted to try the "Long Pool" again, both for legal and economic reasons.

Over many months in 1934 and 1935 majors and independents in California fumbled their way toward a new set of arrangements. Although the form of the arrangements was different, both parties still had to deal with the same old problems, which would not go away and could not be ignored. Town-lot drilling, the law of capture, the race to drain the other fellows' oil from under their leases before they could drain the oil from under yours, and gasoline markets manipulated by the "big dump" service station artists—all left the smaller refiners with no alternative but to seek out the biggest crude oil over-producers (who, if they could drain away their neighbors' oil, could afford to sell it cheaper).

A series of steps gradually brought the situation closer to where it had been. At a meeting of the major refiners in Los Angeles, led by Al Weil of General Petroleum (a Socony-Vacuum subsidiary that later became Mobil), it was decided there could be no agreement to buy surplus gasoline, the "money" crop of the business. Al, a leading California lawyer in his own right, had come up as an independent in the oil business. As a lawyer-oilman, he knew what had to be done and had the courage to do it. As the meeting broke up, he put on his hat, looked over his fellow big shots, and told them that he was not only going to begin buying surplus gasoline, but he thought he could make a profit from it for General Petroleum. With that, he stalked out of the room, and the meeting ended. But it was the beginning of another chapter in the history of the West Coast oil industry.

I saw Al after the meeting, and he grinned at me. "Well," he said, "if we are going to get out of this mess, those of us who are lawyers better do it. At least we ought to be smart enough to know what we can do and what we must not do. I'll need your help." He got it, of course. Over the next few years, I met with Al many times, and we usually saw eye-to-eye. We needed no formal agreements. Sometimes Al had to use his super-salesmanship when gasoline put pressure on the roof of his tanks. When I heard about the pressure, I always packed my sales kit. I remember one conversation with the president of one major company, when I urged him to buy some gasoline from Al. He indicated he could not afford it. I asked him what he thought it might cost him if Al had to dump. That was persuasive.

I once caught one of my associates keeping a little black book, which he called "the reds and the blacks." It listed the amount of gasoline each company had bought compared with what it should have bought if the old agency agreements were still in effect. To his chagrin I took his black book, destroyed it before his eyes, and told him in no uncertain terms that there were no agreements and could be none, but that if someone ever found his book, no one would believe it.

Buying surplus gasoline was only one leg of a structure necessary to reduce the excesses of competition, which threatened bankruptcy for most of the West Coast companies as well as others across the nation. Fortunately, the old Central Committee of California Oil Producers was not abolished with the NRA. The oil code helped, by imposing sanctions for violating the maximum efficient rates of production, which the Oil Umpire's Office set. The committee and the umpire's office, however, were not regarded as violating antitrust laws, either state or federal, and both continued as they had before the oil code ever arrived on the scene. Although certainly not perfect, the committee sought to set rates of production which, over the long run, would produce more for less. Only die-hard classical economists and trustbusters with no knowledge of petroleum engineering and oil reservoir mechanics could really regard reasonable production practices as an unreasonable restraint of trade. The problem now was to find a substitute for the maximum efficient rates that had been calculated by the Central Committee and endorsed by the oil code.

The independent refiners had organized themselves into an Independent Refiners Association. It was really run by a young lawyer, Bill Scully, who had been an associate of mine in Interior before the court abolished our jobs. He too had a pretty good idea of what we could and could not attempt. He let the members of the association know that they were unlikely to sell surplus gasoline to Al Weil if they consistently provided outlets for producers who cocked their wells open regardless of the correlative rights of their neighbors and efficient rates of production in particular oil fields. Those who failed to recognize the facts of life found themselves left to the none-too-tender mercies of the big-dump gasoline service station manipulators. This ultimatum did not fully stop the California version of hot oil, but it took a lot of the profit out of it. That alone put the brakes on the worst of the wasteful and morally reprehensible practices in the oil fields.

Ralph Davies, the imaginative creator of the five-state Pacific Coast Petroleum Agency and Refiners Agreement, looked for a new approach to tame West Coast marketing. He found one in what became known as the Fair Practices Association. As its attorney, I attended all its biweekly meetings, kept its minutes, and prepared its code of fair practices. Its membership included the chief marketing executive of each major oil company and each secondary company, through which the majors marketed a substantial amount (though not the bulk) of their gasoline. The western states covered by the Fair Practices Association were Arizona, Nevada, California, Oregon, and Washington; the territories of Alaska and Hawaii were also included.

The fair practices which the members of the association promoted were based upon the principle that competition could be free and fair only if based upon current, accurate information about the prices of competitors in the marketplace. A number of Supreme Court decisions had defined the assembling of such information as proper under the Sherman Antitrust Act. In addition, the Robinson-Patman Act, which became a part of federal antitrust law in 1936, prohibited price discrimination within the same class of customers, unless it was justified by the lower price of some competitor.

What the Robinson-Patman Act said, in simple terms, was compete but not too much. This approach gave our association permission to try regularly to ascertain exactly what prices were being quoted by each competitor. It took an office and an organization to survey the important metropolitan markets of the West. This was done biweekly for almost three years, and the results were charted and published in books almost as voluminous as telephone directories. These books, which showed not only "posted" prices but the actual price at which gasoline could be purchased at specific locations, took a lot of the rumor, gossip, and hearsay out of the market. The members of the association were personally confronted with the results of the last biweekly survey, and to say that this had a salutary effect was to put it mildly. For good or for ill, it checked a world of deceit in the marketplace about prices.

To be perfectly honest, it probably also led to a lot of pressure and name-calling outside the meeting room between equally devious marketers. I was authorized to stop any discussion at any meeting if I thought it verged on explicit or implicit price fixing between competitors. Sometimes I had to. Since I was sure that someday some zealous prosecutor would accuse us of doing what we were careful not to do—attempting to fix prices—I had a federal court reporter make a verbatim transcript of every meeting. The recorded material included the byplay and bawdy stories, and I was told by some of the FBI agents who did finally investigate us that the stories were the best part of the manuscripts.

Some peripheral incidents were not recorded, however. A vice-president for marketing of one of the companies once said to me outside the meeting: "Howard, we know what we want to do. We want to get these prices fixed. All we want you to do is to tell us a legal way to do it." I told him there was no way, and, even if there had been, it would not work, since few of them trusted each other. The telephone book comparing posted prices to actual prices at hundreds of service stations was regarded by this same vice-president as a nuisance, as he just wanted a list of those selling at "cut" prices. Again, he had to be told that such a list would constitute a black list. What any company did about price cutters was that individual company's business, and none of them could have any agreement, expressed or implied, about it.

Full knowledge of actual market prices did tend to level out the markets. This was shown in the biweekly charts, and, when a grand jury investigation came, these charts were subpoenaed. I refused the originals but supplied copies on heavy photographic paper. The charts were half the length of a fair-sized room, each was rolled tightly, and, after a few weeks, rigor mortis set in. Later, during the war (when I was working alongside and not opposite the antitrust lawyers), I asked them what they ever did with the hundreds of charts. They told me they tried to use them but gave up because whenever they sought to unroll one, it took four people to hold down the corners—then the chart would snap back into a roll like a snake.

After I asked the government lawyers about the charts, they in turn "worked me over." They wanted to know where the "agreement" was—the one that "divided" between companies the surplus gasoline which Al Weil had bought. Apparently they had investigated but could find no pattern as to the quantities each company bought. I told them the truth: they could find no agreement because it had been decided that there could and would be no agreement. I did not add that it would be pretty difficult for them to balance out, for example, exchanges of kerosene in the Far East against gasoline in the five western states covered by the agreement. The tax figures on which the government lawyers relied had to be corrected for a wide variety of intercompany transactions. This was a very difficult job—just how difficult we knew because we did it.

Attendance at the meetings of the Fair Practices Association was mandatory, and tardiness was frowned on. It cost twenty-five dollars to be two minutes late, fifty dollars to be five minutes late, and five hundred dollars to be absent. The fines were assessed and collected, and I always suspected that in some way they landed on the individual's "swindle sheet" (expense account). After the pot was built up over the year, the association spent it on a weekend blow out.

One such party was held in the winter at Yosemite park. Over a long dinner at Ski Lodge overlooking the valley, each executive was presented with a souvenir thought to be typical of his conduct during the year. One received a chisel. My client, Ralph Davies of Standard, got a toy dog with the forepaws bearing the colors of Standard and the tail done up in Signal Oil Company colors. When you wagged the Signal tail, the Standard forepaws jumped up and down. The vice-president for the Texas Company always pointed to his bulging briefcase, where he said he carried the "evidence" of all the wicked things his competitors had done to him. He received a big briefcase embossed in gold with the words, "The Evidence." When he opened it, he found real manure wrapped in tissue paper.

One night, after most had returned from dining to the lodge in the valley, Davies ordered a dozen siphon bottles. He started down the hall with a few of us carrying bottles. He stopped at each marketing executive's room, knocked, and called out, "Western Union." When the occupant opened up, three or four streams hit him. Everyone opened until we got to the room of Hollis Fairchild, Davies's own general sales manager. With all the noise in the hall, he thought it was smart not to respond. But the transom was open, and I kicked off my shoes and climbed on the door knob. I reached back for a siphon bottle, but Davies passed me the fire hose. I pointed it at Fairchild. Instead of merely cracking the valve, Davies opened it wide, and it almost blew Fairchild out of bed before I fell off the doorknob, allowing the fire hose to act like a pinwheel. It soaked everything in the room and the floor below and resulted in quite a breakage bill from the hotel. Some thought we ought to frame that bill under the title, "Oil Men at Play."

Another great marketing party comes to mind—this one at Catalina, off the southern California coast. Some of us wrote a skit depicting a grand jury investigation, and we had a canvas backdrop prepared showing the jury—most of them being asleep. An actual likeness of each marketing executive was attached to a stick so that it could be fitted in the neck of a dummy witness on a stand. The dummy was wired for sound. Each executive had been recorded in his own voice answering "questions," though none knew that the questions would be connected with their "answers." We hired a Hollywood actor to play the prosecutor, and he asked questions of each dummy executive. The recorded answers followed, and it broke up the house. Later, I broke up the records for fear someone might think the answers were true!

We made one bad mistake at the party: we recorded conversations from hidden microphones at each table during dinner. The next night we played back selected passages, and a fist fight broke out. So intense were competitive pressures that one vice-president lost his temper completely when he heard what one of his competitors said about him.

The business of buying surplus gasoline and dealing with marketing practices was quite different from my old activities with the government. I had now embarked on a whole new set of activities directly related to competition and profit; these activities ranged from exploration to marketing. A lawyer for an integrated company will become involved in all aspects of the industry, and there is hardly any better way to learn how the pieces fit together. If you fail to learn the facts in your cases, you are likely to lose most of the time. Personally, I have always hated to lose.

Exploration and production happened to be my first love in the business, and, since everything in oil starts with production, the eighteenth floor of the Standard Oil building, where its senior management was housed, found much for me to do. Although I never served as Standard's representative on the Central Committee of California Oil Producers (sometimes called the Conservation Committee), I was called on to help keep the committee prepared to defend itself, which made me an unofficial member of the committee. More importantly, I became deeply involved in chasing oil and gas leases and acquiring present and prospective producing properties for my client. Many times I wondered whether I was a lawyer in the oil business or an oil man in the law business, as I always seemed to be wearing two hats. Norman Meyers and I had done so for the government from 1933 to 1935, and I have ever since, whether I realized it at the time or not.

Madison Oil Trial

My path crossed with Big Fish's one last time during the long trial of the Madison case. While I was still on the Standard payroll as Special Counsel, the whole oil industry east of the Rockies was indicted for alleged violations of the Sherman Antitrust Act. Among those named in the suit of July, 1936, that challenged the gasoline buying pools (which continued in the last months of the Oil Code and the NRA) were eighteen majors and five subsidiaries, fifty-four of their officers or employees, and three trade journals. Colonel "Wild Bill" Donovan and his firm of Donovan, Leisure, Newton and Lombard, in Washington, D.C., were retained by most of the companies for the defense. Donovan, a former assistant attorney general in charge of the Justice Department's Antitrust Division, was regarded as the leading antitrust lawyer in the country. When my old Yale colleague, Thurman Arnold, left a similar position with the Department of Justice, Donovan's friends insisted that Donovan's firm was about to establish the "Thurman Arnold Memorial Wing" in memory of all the legal business Thurman had brought them.

Standard of California and the Texas Company began to be deeply involved together in the Middle East. Neither of them ever were parties to the so-called Red Line agreements of 1926, under which C. S. Gulbenkian and most of the other major international companies divided up the Middle East. Standard of California, starting with a good discovery on Bahrain Island and expanding with discoveries in Saudi Arabia, found itself with billions of barrels of reserves but no markets. The Texas Company had markets and little crude. A partnership between the two was a natural, and I helped draft some of the contracts for Standard.

The Texas Company and its officers were among those indicted in the Madison case. When the trial began in October, 1937, they asked their partner (Standard of California) to loan me to Colonel Donovan to try to explain the nature of competition in the oil business and to serve as his expert witness on what transpired under the oil code. I was so assigned, and my old associate Norman Meyers was hired as well, for a fairly handsome per diem. His wife and I debated about when he earned it, finally deciding that, since it was a per diem, it must have been when his feet hit the ground getting out of bed each morning.

After a few weeks in Madison, Wisconsin, we persuaded Donovan that he should defend his clients partly on the argument that prices rose because hot oil production stopped, not because of gasoline pooling per se. Certainly unless hot oil output had been stopped, gasoline buying would not have had much effect. (Had some of those who were seriously suffering from past "hot" runs been more patient, prices would have recovered. But it is hard to persuade anyone facing bankruptcy to be patient.) If Donovan were to defend on the grounds that it was the stoppage of hot oil that made the price of gasoline go up, he had to try to prove that there had been massive amounts of hot oil sold. He asked me who could prove it. I told him I knew a real expert on that subject and described F. W. Fischer—"Big Fish"—to him. Could I get him to testify? Answer—"Big Fish" will do almost anything for money. I called him up, made a deal, and he testified. Hot oil flowed in Judge Patrick Stone's courtroom by the millions of barrels. Finally the judge inquired how he knew the hot oil refineries ran these enormous volumes of hot oil. Big Fish put on his best country boy act and replied, "I ought to know, your honor, they was all my clients." After that, even the judge believed him.

But even Big Fish's act was not enough to win this case. A witness from one oil company denied any knowledge of any gasoline buying agreements. He forgot (or later said he did) a telephone conversation with the representatives of another company whose secretary listened in, made a verbatim transcript, and sent ten copies around the company. One copy survived, and the prosecuting attorney read it back to him. It read something like this: "Hello A, this is B. The government is watching us very closely, so I am telephoning rather than writing. We are going to start buying gasoline on the Group III market in the morning. We expect to start at such-and-such a price and raise it in increments of an eighth of a cent per gallon until we buy such-and-such an amount. Your share is so much, do you agree?" The recipient of the call agreed. After that, it was hard to claim there were no agreements. The leading attorney for the government had a yellow chart portraying the regular, ladder-like advances of gasoline on the Group III market. Again and again, he walked to the chart, used a pointer, and looked at the jury with the words, "Now let us once again climb these golden stairs of greed and avarice." He was so effective that, when he left the government, he became a partner of a leading law firm representing one of the major defendants in the Madison case. If you can't lick him, hire him—a time-honored method for legal advancement.

Competition between big companies, and even between big partners, is so intense that all alliances are extremely fragile. Here I was on loan from California Standard to help the Texas Company that, with only 4 percent the gasoline market on the West Coast, broke the whole market in the very area where Standard had many times this percentage. This made Standard furious. A call from the president of Standard ordered me to come home as soon as I told the Texas people why I was leaving, and I took the train for San Francisco. There I was served a subpoena from Donovan and the Texas Company to testify in the Madison case. I had to face the competition again, and soon I was back in Madison. I informed the Texas Company (and my friend the Colonel) that there was no way an expert could be compelled to have an "opinion" and that as far as my government activities were concerned, they were "privileged information"—matters relating to an attorney and his client, the secretary of the Interior. The Colonel, a fine person, grinned. I had made my case. A few hours later, I was on my way back home for good.

On June 2, 1938, most of the defendants either got dismissed or paid relatively nominal fines. The prosecutors finally got their pound of flesh with the conviction of a few sacrificial lambs, and Supreme Court appeals would take the case into 1940 and 1941. As usual the lawyers were the only real winners, and the prosecutors got good legal positions in private life. Donovan's law firm sent a group of the oil companies a bill for some six million dollars. As one of the convicted defendants remarked with a wry smile, "I wonder what he would have charged if he had won." The case, brought several years after the alleged illegal acts, had no measurable effect, for good or ill, upon either the oil industry or the general public. As William O. Douglas said later, "There are those who believe we can spend our way to prosperity. Thurman Arnold with his trustbusters wants to try to sue our way to prosperity." The public really pays for the lawsuits—in millions of dollars of legal fees and expenses either charged to the taxpayers directly by the government or passed on indirectly by the defendants in the cost of the products sold to consumers.

Long Beach Oil Development Company

One of my most interesting production assignments revolved around the Long Beach Oil Development Company, still known as the LBOD. In December, 1936, the General Petroleum Corporation discovered the Wilmington oil field just off the flank of Signal Hill. It became the nation's second largest oil field with over a billion barrels of reserves located in pay sands between 2,200 and 5,900 feet under approximately 3,500 acres. It covered town lots within the Long Beach City limits, most of Long Beach Harbor, and properties of the Union Pacific Railroad, the Ford Motor Company (where the first well was drilled), the Craig Shipyard, and a great number of smaller interests. No surface structure marked the field. General Petroleum, which located the field with geophysics, initially had very little idea of its magnitude either by surface area or by the depth and multiplicity of its many prolific producing horizons—you never really know until many wells are drilled both laterally and vertically. Exploration is anything but an exact science, even today.

Because of its location and the scattered cut-up character of its surface ownership, Wilmington touched off another wild town lot drilling campaign in the Los Angeles Basin. If anyone had told me in

Twelve wells spudding in a section of the Wilmington field for the Long Beach Oil Development Company.

advance that this could happen in a state with no conservation laws without wasting most of the potential reserves of the field, I wouldn't have believed it. But it did happen, and one of the keys to it was the LBOD. There were other keys, however, and these will be explained in due course.

The Long Beach Harbor lands originally belonged to the State of California. They were assigned to the city "in trust" for harbor purposes without the apparent authority to sell or lease. Nevertheless, the city put up these lands to competitive bid for oil and gas leases. A group of Oklahoma oil companies—gambling promoters but not much else—bid them in. These leases were as proven for big reserves as any I had ever seen, and some of us could picture what would happen if the leases were awarded to the Oklahoma promoters. Simply stated, since there existed an almost overwhelming surplus of oil already, all hell would break loose in the market.

What should we do? We invented what for years was known as the production sharing contract and was copied worldwide in almost every important international contract for several decades. I even borrowed my own invention several times in my international oil dealings in Mexico, Argentina, and Indonesia. For the historical record, it started at Wilmington, California, born, as was the Federal Tender Board (of hot oil fame), out of sheer desperation. Necessity again was the mother of invention.

It was simple after we thought of it. I learned at the Yale Law School that the mere legal title to something meant little; the benefits that flow from title are what matter. So why worry about title? If the city retained its title to the harbor lands and merely contracted with an oil company (or companies) to drill and develop its lands for a fee, a doubtful legal question would be solved. If that fee was paid in money, measured by a percentage of the produced oil, it was still a fee and not an assignment of title. The city fathers and the Long Beach Harbor Board were persuaded by this plan. They rejected the bids of the Oklahoma promoters and readvertised for new bids on the basis we had suggested.

The task of developing the Long Beach Harbor lands seemed too big for one company—even the largest. This was particularly true since one of the conditions of the bid included a guarantee to buy all the oil which might be produced from the harbor lands at the average posted price of the major purchasing companies in the area for oil of like grade and quality. Even Standard of California feared this condition. Once Standard posted its prices, it constituted an offer to buy all oil offered at such prices.

The LBOD was organized for the bid. Most of the so-called major companies originally subscribed to its shares, together with some of the important independents, notably the Signal Oil and Gas Company and the Hancock Oil Company. Just before the bids were filed, most of the majors backed out because of their concern over the oil buying commitment. Harry Sinclair, well past his prime, ordered his old friend and subordinate, Richfield's Charlie Jones, who had incorporated the LBOD, to withdraw. Charlie gave me the news with real tears in his eyes. This left only Standard, Signal, and Hancock. Somehow we helped hold Standard firm, perhaps because I argued with Bill Berg, the president of Standard, that whatever the volume of oil we might have to buy, the contract would constitute an option to buy oil at a little less than our competitors. If the volume got beyond what our refiners could run, we could always drop some other oil. The economic argument should have won on its own merits but did not. I won only because Berg was an honorable man, and he finally agreed to go forward because, as he put it, "Standard gave its word. We always try to keep it."

After the LBOD won the bid, Will Reid, the head of Hancock Oil, told me his one-third commitment was more than he could handle. He asked if Standard would take part of his purchase obligation in return for an equal amount of Hancock's interest. Again I argued its economic merits with Bill Berg. I got a quick answer. Bill told me to tell Will that Standard had lived up to its word, and Standard expected Hancock to do likewise. Hancock did and bought its share of a large amount of production. World War II was just around the corner, however, and the obligation to buy carried with it an obligation on the part of the city to sell. Therefore, it became an asset, not a liability, for Hancock. My economic arguments with Bill Berg turned out more correct than I had dreamed. I was right again for the wrong reason!

I still remember the form of the winning bid. LBOD agreed to take all the risk and put up all the money to develop the city properties. For this service, the company was entitled to a fee of 13.44 percent of the value of all the oil which was produced from the lands of the city for twenty years. From the value of the remaining oil, not quite 87 percent, the company could reimburse itself for all of its development and production costs. This left the city with the value of all the oil but the 13.44 percent fee after the company retired its development costs and operating expenses. The city retained title to the oil at all times, until the company sold the oil and divided the proceeds in accordance with the terms of the contract.

What we had done was to reverse the normal position of a royalty owner and operator. The city (as the landowner) paid the costs of development and operations, and the company (as the operator) received only a fixed percentage of the value of the oil—cost free and tax paid. Such an arrangement was unheard of at the time. Even those of us who proposed it would not have had the nerve if we had not regarded the city lands as proven for millions of barrels of oil. This minimized the risk for the operator.

There was much more in the contract, which ran nearly five hundred pages. The term was twenty-four years and nine months, set to expire in February, 1964.

There was one risk that we had to take. The harbor lands were offset with a line of twelve producing wells drilled by other oil companies on one-acre plots—a distance considered conservative spacing for a town lot area in the Los Angeles Basin. The contract with the city required that each of these twelve wells be offset by drilling a well within thirty days of the contract. If we failed to meet this time schedule, we lost the contract and whatever had been spent up to that point. During the bidding stage, I served for the first time as a vice-president of an oil company. I had expected to resign when the formal procedures were finished. My friend Tex Lombardi of Standard told me to stay on in an executive position. I did, worrying about getting those twelve wells started from swampland drill sites within thirty days. I still have in my office a picture of the twelve wells taken on the twenty-ninth day, when the twelfth offset was spudded. The picture shows an oil field as they once appeared—tall, steel derricks and big, steam-driven draw works (the drilling machinery). The picture does not show Ernie Pyles of Hancock and Howard Marshall of Standard baptizing the start of the first well after an evening of scotch and soda at the Pacific Coast Club. Nor does it show the log of that first well, indicating hundreds of feet of saturated oil sands in a whole series of separate horizons.

On the day when we expected the hammer to fall in our favor as the high bidder, we met with the Long Beach Harbor Board. We got wind the evening before that some of the unsuccessful bidders were going to appear at the meeting to charge that the LBOD was nothing but a "corporate shell." Will Reid of Hancock asked whether I could get Standard to have one million dollars on deposit early the next morning, crediting Long Beach Development that amount and dividing it between the four largest banks in Long Beach. I talked to the treasurer of Standard late in the evening, long after the banks were closed. I never found out how he did it, but four passbooks for a quarter of a million each showed up in my hotel room before seven the next morning. When the expected charge of a "mere corporate shell" was leveled at the Harbor meeting a few hours later, I threw down the passbooks on the board table. I suspect few of the board members had ever seen passbooks totalling a million dollars. At any rate, it silenced the opposition, and the award was made. Naturally we drew the money out the next day and replaced it with a lot more as needed. After all, it was only a dramatic gesture in the first place— worthy of Big Fish at his best (or worst, depending on which side you were on). Instead of the "poor country" lawyer act we put on the "big company" act.

The State Lands Act of 1938

In Madison, we were dealing with history. Back in California, we faced the present and the immediate future. One of our concerns revolved around the Huntington Beach oil field. Ever since the slant-hole artists solved the technical problems of drilling directional holes, a euphemism for what many called "crooked holes," Huntington Beach had become their favorite playground. Dozens of wells were started from locations onshore and bottomed out under the Pacific Ocean in the state-owned tidelands. None of these operators bothered to get a lease from the State of California or took the trouble to acquire crossover permits from intervening landowners. Such activities involved nothing less than deliberate and willful trespass to steal the state's oil. As might be expected, only fly-by-night individuals or questionable oil companies indulged in these activities. Legitimate operators rightly feared the consequences of playing this kind of hide-and-seek.

As with most illegitimate practices, efforts were made to make them appear justified. The political approach is frequently a convenient refuge for scoundrels, and, at Huntington Beach, they were quick to try to use it. They claimed that the major companies were trying to put them out of business by pointing out to state authorities what was going on. Others told these same authorities that they did not know where the wells were bottomed, but their favorite ploy was to accuse poor Standard Oil of being the real criminal.

The accusation came about this way. In the mid-1920s, Standard took an oil lease from the Pacific Electric Railroad on a long strip of land the railroad owned between the Pacific Coast Highway and the Pacific Ocean. In those days no one knew for sure where the bottom of any well really ended up. It was not until the early thirties that the engineers perfected an application of the gyroscopic compass, which was unaffected by the steel pipe in the hole, that would give a true north-and-south reading of the bottom of a well together with the horizontal deviation. Before that, we knew well bores wandered vertically, but not where. Standard had drilled a line of wells on the Pacific Electric strip.

To divert attention from their own illegal activities, the crooked holers pointed fingers at Standard, shouting that some of Standard's wells were bottomed out in the tidelands. Standard had its wells surveyed, and relatively few were found to have drifted a few feet out from under the surface leased from the railroad. The turmoil and the shouting were heard as far away as Sacramento. The deliberate stealing of the fly-by-nights, running into the tens of millions of barrels, was temporarily ignored. Attention instead concentrated on the relatively minor amounts of oil produced by Standard from a few wells which had inadvertently drifted a few feet into the state tidelands. Moreover, the crooked hole operators argued that they were protecting the state because Standard had been "draining" the state tidelands even from the straight holes drilled on the railroad lease. This was probably true—another example of the law of capture long ago legitimized by the courts. Naturally those who had deliberately bottomed their wells under state property failed to mention that they were paying the state nothing for the oil they were draining. It was hot oil all over again but only "hot" because the operators did not pay anybody anything.

Bill Berg, now president of Standard, jokingly offered Ralph Davies, his marketing vice-president, and me the best dinner we could eat if we could figure out some way of solving the Huntington Beach mess. Previously, an attempt had been made to persuade the legislature to lease the state's tidelands to protect the state's interest and stop the stealing. These attempts had always failed because of warring factions among the independent oil companies, each seeking a favored position to acquire development rights in the offshore extension of the Huntington Beach oil field. So vicious were the battles, that arguments over one proposed piece of legislation ended up in a fist fight between two prominent independent oil executives on the overnight train from Los Angeles to Sacramento. Bad blood was literally spilled.

A new start had to be made. Davies and I put together a group, the Southwest Exploration Company, to sponsor corrective legislation. It was initially owned in equal shares by Signal Oil and Gas, the Hancock Oil Company (both integrated independents), and Standard. I journeyed south to meet with Ed Pauley and John Elliott, both independent oilmen and leaders of the Democratic party in California. It was widely believed that they had the ear of Gov. Culbert Olson, who was also a Democrat. Pauley and Elliott, heretofore arch enemies of various legislative attempts to deal with the tidelands, were offered 10 percent of Southwest Exploration if we could devise and pass legislation which would permit the state to lease its offshore properties at Huntington Beach. They could divide up the 10 percent with other independents who could help. Pauley and Elliott agreed, and each of them took 4 percent, further dividing the remaining 2 percent. While the percentages may have seemed small, the dollars to be made were quite respectable when you multiplied them times hundreds of millions of barrels of proved oil reserves. No one got a free ride, as all the owners of Southwest had to defray their share of the cost of acquisition and development.

The legislative struggles resulted in the passage of what became known as the State Lands Act of 1938. Long and nasty were the words thrown around in the halls of the legislature. I was sitting in the gallery with Harry March, vice-president and general counsel for Signal, when we were subjected to a particularly vitriolic attack. I asked Harry what he thought it would take to answer, and he replied, "Forty-one votes and I think we have them." When the roll was called, we had forty-three. Harry knew his politics—I was just learning.

We even persuaded Ralph Davies to join us in buttonholing legislators to ask for their support of the bill. A vice-president of Standard talking directly to state assemblymen was almost unheard of. One evening, sitting in the lobby of the Senator Hotel in Sacramento, Ralph and I heard two legislators talking about a bill with which neither of us was familiar. One legislator dropped his voice to whisper, "You know who is behind the bill?" The other asked, "Who?" "The Standard Oil Company," replied the first legislator. The other replied, "Those smart people, they think of everything." Ralph turned to me with a laugh, "Don't you wish we were that smart?"

The State Lands Act of 1938 was a good bill. Clearly it was in the public interest, as it empowered the State Land Commission to lease the tidelands to the highest bidder. Fortunately for us, it did not permit the state to condemn private upland drill sites from which directional wells could be drilled. It prohibited the drilling of wells in the ocean from piers or artificial islands, and this was at the behest of the Save the Beaches Association, a forerunner to the environmental groups that regularly protest offshore drilling in California today. In this case, however, it was rumored that certain oil interests helped finance the association, since these same interests were thought to control the upland drill sites. Personally, I had reason to believe the rumors might be true. If so, it would not be the first time (or the last) when public and private interests joined hands for different reasons.

Independent oil interests that owned no part of Southwest Exploration opposed the land act bill. They assumed that their friends, Ed Pauley and John Elliott, would get the governor to veto the bill. The deal that had been cut in Southwest Exploration had not been advertised. To the amazement of the opposition, the governor signed the bill. We prepared to defend against a referendum to defeat the legislation, but some of our friends headed the referendum campaign. They did an excellent job by failing to get enough signatures to put the question on the ballot. It was said that enough signatures had been secured, but somebody broke into the office where they were kept and stole a batch of the petitions the night before they had to be filed. At any rate, there was no referendum, and we all breathed a sigh of relief. Such was the rough-and-ready character of California politics at the time.

It had been a long battle—almost two years. The lands—835 acres worth—were leased, and Southwest was the high bidder. It alone was able to put together the best drill sites, although I negotiated with my client and Pacific Electric to acquire the legal right to drill through Pacific Electric's strip, which fronted on the ocean. Southwest bid the highest royalty rates to the state based on a formula we invented—the bigger the well, the higher the royalty percentage, limited by maximum efficient production rates. It taught me one lesson which all experienced bidders know: if you help write the conditions of the bid, you can frequently out-bid the competition.

One can drive along the Pacific Coast Highway today and see the results of our efforts. Nowhere else in the world will you find two miles of oil wells, bumper to bumper and two lines deep. Actually, they only appear to be spaced this closely. The wells are all bottomed out under the ocean over an area several miles long and almost three miles deep.

The course of the wells was surveyed and charted, and a map showing their courses and bottom hole location looks like a bowl of spaghetti. The wells were directionally drilled to hit a target as far as three miles from their surface location. So accurate was the engineering that each well was kept in a fifty-foot cylinder all the way down.

The layman may wonder how all this was done. It is simple enough: the drill pipe which turns the bit is limber when you join hundreds and thousands of feet together. After setting a straight hole for a few hundred feet, set a "whip stock" (which is nothing more than a steel wedge) to point the bit in the desired direction. Use an oil-well surveying instrument at regular intervals to identify your location as you drill. More weight on the drill bit will cause the well to drift further from the vertical; less weight tends to bring it back toward the vertical. Sometimes when we entered the oil sands laterally (far removed from our starting point), the well would be 70 percent off the vertical. Some of our engineers used to say they almost had to put jacks on the steel pipe (with which the completed well is cased) in order to force it down the hole. Normally it just drops easily (sometimes too easily) with the pull of gravity.

Reservoir engineers think of their art as a more or less exact science, though often it seems to be less rather than more. Many wells were drilled at Huntington Beach after the tidelands issue was put to rest. All these wells were carefully logged as they were drilled, and we thought we knew all about the field. But it took a decade before we stumbled onto the fact that a series of upper horizons (through which we had drilled for years) actually contained vast amounts of relatively low-gravity oil. Perhaps a correct reading of the electric log should have told us, but we missed it entirely. We learned only when one of our engineers thought he might as well test one of these upper zones. There it was and had been all the time.

Sometimes we thought we had a nondepletable oil field. At the end of each year, when the engineers reappraised our reserves, we seemed to end up with as much or more than we'd thought we had at the start of the year, despite the production of millions of barrels. Equally competent and honest reservoir engineers will vary by wide percentages in their reserve estimates, each using exactly the same data. This makes for an interesting horse race, as most in the oil business know. Do you buy on a seller's appraisal or sell on a buyer's appraisal?

As I have driven past those two miles of wells at Huntington Beach, I remember the years it took to acquire the leasehold rights to develop a known oil field. Yet it took only a year or so—beginning with the first well that was completed on November 19, 1938—to drill the wells to tap the so-called main zone in that field. As so often happens, the physical problems were simple. The political and economic problems cost blood, sweat, tears, and time. Things aren't much different today. Finding and producing adequate supplies of hydrocarbons in general, and oil and gas in particular, pose no serious physical problems. Our so-called shortages in the 1970s arose because decision makers in Washington, D.C., were unable or unwilling (probably both) to recognize that shortages come from an inability to overcome economic and political barriers.

Once during the war I testified before a Senate Committee. A senator asked, "Mr. Marshall, are you going to run out of oil?" I told the senator he had asked the wrong question. He then inquired what he should have asked. "You should have asked me whether we were going to run out of hydrocarbons. The answer is no. It is solely a question of what it will cost to produce them and whether you will allow the producer, public or private, to recover those costs." I knew that more oil and gas than we had thus far discovered lay beneath the lands and the offshore provinces of the United States; but that it would never again be produced for less than the dollar a barrel it once had cost to find and produce these resources.

The Huntington Beach mess was finally laid to rest with our new law. Ralph Davies and I called the bet that our president had made—the best dinner we could eat at his expense. We requested Perino's Restaurant in Los Angeles—as good as Maxim's in Paris and about as expensive. We brought along half a dozen of those who had fought the battles with us. We knew our president never carried much cash, so Ralph conspired with the owner of the restaurant, Alex Perino, to run up a bill that Bill Berg would be unable to cover with whatever cash he might have with him. Perino gave his enthusiastic cooperation. We started with cocktails and a liter of fresh Iranian caviar. The steaks were out of this world, the wines of the same quality, the brandy hundred-year-old Napoleon. When the check arrived, we all watched Berg. After he recovered from the shock, he walked over to the cashier to pay with a check. He returned with the cashier, who demanded to know whether this man really was the president of Standard of California. We greeted this question with gales of laughter, and Bill probably thought he was going to have to wash dishes before he could leave the restaurant. We eventually confirmed Berg's identity. He enjoyed the joke—afterward!

Ralph and I regularly ate at Perino's when we were in Los Angeles, but there was a little game Ralph had to play. It became a routine to leave the restaurant and see just how close we could come to not missing the Lark for San Francisco at Glendale. If we arrived just as the train was starting to move, it was par for the course. If we arrived a minute earlier, Ralph always complained that here we were standing around stations and airports again. If we arrived minutes late, we chased the train all the way to Ventura or Santa Barbara in a big twelve-cylinder Packard driven by Standard's best chauffeur, a former race track driver. One night, we had to go all the way to San Luis Obispo to overtake the Lark.

We all worked hard in those days, but we played hard too. Once, we took Winthrop Rockefeller, grandson of John D. Rockefeller, on a cruise on Signal's 120-foot yacht and got him rigged up for swordfishing. After he fished for a bit, Sam Mosher, head of Signal, pretended to find something wrong with the tip of the rod. While "fixing" it, he attached a subsidiary line to the main line and hooked it to a five-gallon bucket. When the bucket was furtively tossed over the side, Winthrop got a strike he could not hold. Out went a few hundred yards of line off his reel, and the captain slowed the boat to allow him, after much physical effort, to recover most of the line. After an hour or so of playing the line, young Rockefeller was allowed to catch his "fish." We told him that in California this was known as catching a "buckeeta."

Ralph's practical jokes on others were seldom turned back on him. But Frank Buck, the head of the Golden State Milk Company, once managed the trick. Returning from Lake Tahoe, Ralph and I stopped at an artist's roadside stand. Ralph bought a perfectly horrible oil painting of Lake Tahoe. He had the art department of Standard erase the artist's name and forge a famous one. It was presented to Buck with a note to the effect that while it could not compare with the Corots hanging in Frank's office, it still seemed to catch the spirit of Lake Tahoe. A puzzled Frank said nothing. A few days later the monstrosity was hung between the Corots with a sign: "Personally presented by Ralph K. Davies." Ralph stood it for a week before he threw in the towel by telling Frank, "I give in—for God's sake take it down."

Our problems in California did not evaporate overnight. The glue that sometimes held the situation together was far weaker than some of the "wonder glues" now advertised on television. With the State Lands Act of 1938 (and the final leasing of the Huntington Beach tidelands), perhaps one prospective disaster had been averted. But there were plenty more on the horizon.

From Standard to Pillsbury

I had been an employee of Standard of California for about two years when a slight career change occurred. Standard's outside counsel, Pillsbury, Madison and Sutro, had an office just a floor above their major client. I was doing legal work for Standard as Special Counsel and was in close association with the firm. After a reorganization, Ken Kingsbury, president of Standard, asked if I would mind leaving Standard to be with Pillsbury. I thought it was a good idea. He then invited Alfred Sutro, one of the firm's principals who had retired in 1929, to join Standard as vice president and chief counsel. Sutro consented, and the deal was struck with Pillsbury that I would have a year with the firm and would join permanently as a partner if all went well. This was fine with me—I would remain immersed in petroleum legal matters, which was my calling. If things did not work out, I would rejoin Standard.

A year later, I thought everything had gone well. Yet I heard nothing from Sutro, and my salary remained less than when I left Standard. I called Kingsbury about it, and he, as I learned later, immediately called Alfred. "Alfred," said Kingsbury, "wasn't there an understanding that Mr. Marshall would become a partner after a year if all went well?" Before Alfred could respond, Ken added: "Everything has gone well, hasn't it?" Alfred replied that he knew of no problems and that I had done "very well." If Sutro didn't know the reason for the call, he did after Kingsbury's next statement: "Alfred, you know that Standard does not like its legal business handled by clerks."

Several days later, Alfred called me into his big corner office and, with every pleasantry, told me that things were going swimmingly. He had talked to the firm, and I was invited to join as a full partner. After a chuckle that I could not suppress, I said I would be delighted. I knew what had jogged his memory.

STANDARD OF CALIFORNIA
Long Beach Turning Basin

Another leasing challenge for Standard was the Long Beach Turning Basin, some seventy to eighty acres of water-covered oil lands in the heart of the Wilmington oil field. The title was cloudy. Some character, claiming title, offered a lease to Standard. Standard turned him down on the grounds that it did not drill town lots. I remarked to Standard's vice-president of production, Tex Lombardi, that even town lots with hundreds of thousands of barrels per acre could still produce millions of barrels of oil. Did he have any objection to my taking the lease over to our friends in Signal? At least if they made a deal and were successful, Standard would have a chance to buy the oil because of our contracts with Signal—as opposed to more dumping on an already chaotic market. I got permission to work with Signal.

Sam Mosher of Signal jumped at it on one condition—Standard had to allow me to work with Signal on the title questions and legal problems. It took months, but we finally bought out the adverse claimants and got the lease, and Signal drilled the whole series of wells into the turning basin. I can still see Signal vice-president Garth Young watching his company drive the first piles for a drilling location. He started to cry because the piles were some seventy-five feet long and cost seventy-five dollars apiece. As the driver hit the first pile, it dropped twenty-five feet. Garth moaned, "Twenty-five dollars for each hit—I can't bear to watch it."

One happy ending that made me some dollars with little risk came from the turning basin. Sam Mosher asked my client and my law firm if he could give me a hundred shares of Signal to thank me for bringing him the original proposed lease and helping him solve the legal problems. They both consented, and I took the shares as a "fee" rather than as a gift and paid income tax on their market value of twenty-six hundred dollars. I still own the shares (now worth several hundred thousand dollars). Like the Signal shares I bought after Signal and the other companies were awarded the Huntington Beach tideland leases, Sam Mosher claimed they were "too good for the common people." My wife described my conduct differently. She insisted I was a packrat, always buying and never selling.

I almost violated my own packrat rule in the case of the Signal shares I bought after the award of the tideland leases. One morning I came out of the State Lands Commission with the lease in my pocket

The scene of some practical jokes—Signal's yacht. *Back row, left to right:* Ralph Davies, John Black, Oliver Lamson, Sam Mosher; *front:* Marshall and Winthrop Rockefeller. *Courtesy Ralph Davies family*

and realized I knew something that was not yet generally known. I called Ralph Davies on the phone with a simple question. "I have the lease. Standard has millions of shares outstanding, Signal only 160,000 shares. There are at least several hundred million barrels of oil under this lease. It will have little effect on the value of Standard, but Signal's third of those barrels could make the company. You are my client. Is it all right if I buy a little Signal?" With Ralph's permission I bought 500 shares, fully intending to sell them quickly before I had to pay for them. We went on another vacation in the Gulf to celebrate, but I was counting my money much too soon. While on the trip, the Munich Conference almost precipitated World War II ahead of time, and stock markets plunged everywhere. Signal went down, not up. I put a mortgage on my house and paid for the stock, some sixteen thousand dollars. Today that stock has a market value of well over a million dollars, and dividends have repaid my investment many times over. All of which proves, as I have long known, you do not always have to be smart to make a dollar—sometimes stupidity, luck, and good timing do better. As DeGolyer once said, "In the finding of oil, it's good to be good, but it's better to be lucky."

Bolsa Chica Gun Club Lease

Another potential disaster point—in terms of too much oil being discovered—was again located at Huntington Beach. In 1920, Standard had leased the lands, partly covered with water, of the Bolsa Chica Gun Club, which was owned by a group of well-to-do sportsmen. A series of prolific wells offsetting the duck ponds established the productivity of the gun club properties. Unlike most oil leases, Standard's gun club leases were for a fixed term of twenty years, rather than for as long as commercial production continued. Standard, however, could hold a lease after twenty years for a limited area around each well that might still be producing at the end of the term. Although Standard originally held a lease on the duck ponds, the contract specified that it could drill only from the uplands so as not to disturb the hunting and fishing activities. Before the art of directional drilling had been perfected, there was no way to get at the oil obviously in place under the ponds.

Standard's twenty-year leases were set to expire on July 1, 1940. I suggested to Bill Berg, the president, that we quickly drill enough directional wells to hold our Bolsa lease. Berg refused on the ground that Standard did not drill crooked holes. Besides, as he put it, Mr. Valentine, the head of the Gun Club, would take care of us. I argued that directional holes were perfectly legal if you owned a lease on the minerals where the wells were bottomed and, in this case, we presently owned a lease on the minerals under the duck ponds. Moreover, it seemed to me that Mr. Valentine was likely to try to take care of himself, since he (together with all the members of the club) individually owned their share of the minerals under the club lands. (This was, again, an unusual situation.) My worst fears were realized when, the moment Standard's lease expired, Bolsa offered for competitive bids a new lease on the duck ponds for oil and gas development. Obviously the ducks were no longer important compared with the tens of millions of barrels of proven oil production.

Those millions of barrels constituted a new threat to an already loaded market. Traveling one evening on the Lark—the Southern Pacific night train to Los Angeles—I sat in the club car with Ralph Davies, Standard's marketing vice-president. Ralph was fond of saying that the Lark was appropriately named and fortunately crossed no state lines. We agreed that if another piece of Huntington Beach broke loose, the results would be no lark. We also agreed that our own production people were far too conservative to ever compete effectively for proven oil under Bolsa. Since they already had more oil than they knew what to do with, they might kid themselves that the duck ponds had already been drained. Rather than sit out a losing battle, we decided to talk to our Signal friends once again. Bolsa copied the idea of the production-sharing contract, which here took the form of a "net" lease as opposed to the normal "gross" royalty. They asked for bids; in simple terms, they asked operators to bid on the basis of how much net profit the operator would leave with Bolsa after all costs were recovered.

Signal won the competition by a very small fraction of a percent. Bill Whiteford, later head of Gulf, then headed British American Oil, the next highest bidder. Years later he told me he thought he had had it wrapped up. He wanted to know how we had bid it. Since Bill liked to play bridge, I told him that one peek was worth two finesses.

As Ralph and I expected, Standard was not within shouting distance of the high bids. Amusingly enough, the first wells were drilled for about half of Standard's estimated costs and came in flowing many times more oil than our engineers had projected. But if we had lost the production battle, we had once again won the market war. Signal produced its wells in accordance with good conservation practices, and Standard had the right (and obligation) to buy the oil.

Signal offered us a 50 percent participation in the successful bid. For reasons I did not understand then or now, we turned it down. At about the same time I was asked by Bill Berg to explore with Sam Mosher whether he would be willing to take over Standard's half-interest in the Southwest Exploration Company, which was then developing the Huntington Beach tidelands. All Bill wanted was forgiveness for a few hundred thousand dollars of a receivable Standard owed Signal. I told Bill I wished I could buy it myself at that figure. When I laid it before Sam Mosher, he said simply, "You can't be serious." As I feared, he would hardly let me out of his office until the contract was signed and sealed.

Later I told Ralph Davies I was worried about our future, as I had watched us relinquish hundreds of millions of barrels of oil reserves at Wilmington, the Turning Basin, Southwest Exploration, and the Bolsa Gun Club. He wisely told me, "Don't worry—there is so much fat around here from the efforts of our predecessors that no one will ever notice the difference," and he was right. As an employee at a big company, your greatest risk arises from action that fails—not from inaction that wastes great opportunities. Ralph used to keep a turtle on his desk bearing a sign which read, "Behold the turtle, he only makes progress if he sticks his neck out." He followed that maxim, and it was, perhaps, that attitude which cost him his association with Standard later. Innovators are never popular with those who believe they have it made.

Fighting for Proration

After winning so many where we stuck our necks out, we were about to lose a critical struggle. We knew at the outset that much—even our future with Standard—was on the line. Though we had made an effort to hold the line against wide-open, wasteful rates of production from the derrick forests which marked the overdrilled California oilfields, the maximum efficient rates of production determined by the Central Committee of California Oil Producers were only "guidelines." Like many another "voluntary" efforts to mitigate the wasteful excesses of competition driven by the law of capture, such guidelines were weak reeds to lean upon. The answer to the problem of overdrilling had to be a California conservation law modeled on the best of the mid-continent statutes, which had already proved their worth. I think I could have drafted one in my sleep, but I didn't dare. Politics being politics, it would have been inevitably tracked back to me if I had done so, and my association with Standard Oil would have killed it, regardless of its merits.

Ralph and I, however, entered upon a bit of "advanced connivery." From my Washington days, I knew the head of the Oil Workers Union, and, I thought that if I could persuade him to initiate and sponsor an effort to adopt conservation legislation in California, no one would be likely to suspect we were one of the prime movers—particularly since Standard was widely regarded as anti-union. Such legislation would benefit both the people of California and the oil workers. (Gasoline shortages in California in the 1970s established this fact, if there was ever any doubt.) I had the New Mexico statute, circa 1935, copied verbatim.

Key sections of the bill prohibited crude oil production in excess of "reasonable market demand" and established a state Oil Conservation Commission. The head of the Oil Workers, Harvey Fleming, handed it to Sam Yorty, later mayor of Los Angeles, and a young California assemblyman from Long Beach named Maurice Atkinson. They embodied it into a bill, and after a bitter legislative fight, the Atkinson bill passed, and the governor signed it on July 2, 1939. We had had many of the same crew working for its passage as we had for the State Lands Act of 1938.

But, alas, in California any law is subject to a referendum. All you need to get is a relatively small percentage of the total number who voted in the last election to sign a petition requesting a popular vote on the law in question at the next election. In those days, it seemed, you could get the required number of signatures on anything for ten cents a name. If you were willing to pay for it, a number of so-called public relations firms would guarantee it. People on street corners sign anything if the issue is properly "explained" to them. All you needed was enough solicitors and enough street corners. Our opposition promptly put up the necessary funds, and poor old AB1926 was scheduled for a "yes" or "no" vote at the next general election.

Our opposition did not spontaneously arise from some ground swell of public opinion in the state. For those who believe the oil industry is a monolithic monopoly, the inside story of the referendum campaign would be an eye opener. As usual, the oil industry was hopelessly divided over mandatory proration. It didn't even split on major and independent lines, as some of the larger majors, like General Petroleum, supported the new law, while others, like Union Oil of California, opposed it. The integrated independents were likewise split, with Signal and Mohawk for and Hancock against the referendum. Non-integrated producers believed, probably correctly, that the law of capture was to their financial advantage. They, and others, poured forth the funds to wage an advertising and political campaign against the law. Those favoring the law did likewise.

Some years before, a similar conservation statute known as the "Sharkey" bill—so named after its author—was passed. It was beaten in a referendum in May, 1931. Billboards and cartoons from that old campaign were resurrected by the opposition group, the Independent Petroleum and Consumers Association. They featured an enormous shark, of course labeled Standard Oil, about to swallow a whole school of frightened little fishes, labeled "the independents." Naturally, no mention was made of the size of those financing the opposition to the tune of $400,000. Certainly Superior Oil and Union of California, to take two examples, were anything but little fishes!

The campaign took me out of legal circulation for several months as I stumped the northern part of the state, day and night. I used all my political muscle by asking FDR, former president Herbert Hoover, Interior Secretary Ickes, and California Governor Culbert Olson to support the new law. Despite all efforts, however, we lost. The total vote was 39 percent for and 61 percent against. Victory in the north didn't save us from overwhelming defeat in the south.

Winning any referendum is a difficult task. You start with one-third of the votes against you: those who do not understand a bill automatically vote no. This alone means that you have to convince two-thirds to vote affirmative, which is usually an almost impossible task.

At the end of the campaign, I had to struggle with our financial housekeeping. The state laws required that monies raised and spent in any campaign be reported once on an interim basis during the campaign and then at the conclusion. We filed the interim report showing relatively minor expenditures. When it was all over, we figured no one would worry much about the losers' final report, which, incidentally, accounted for well over a million dollars. As I recall, a full report was prepared and placed in the mailbox. To be more precise, it was put—but not dropped—in the mailbox. Needless to say, it never made it to Sacramento. We never saw the victor's final report either, but we surely felt the blows from what they spent! Besides, we knew pretty much where their funds originated, just as I suspect they knew about ours—there are no secrets in the oil business that remain secret for long.

Far more serious than political defeat was the fact that California once again had "cut off its own nose." How much of its cheap oil had been left in the ground, how much of its natural gas blown to the air—no one knows. No one can deny that the losses run into the hundreds of millions of barrels of oil and billions of cubic feet of gas. When the lack of even elementary conservation practices in the older oilfields is compounded with arbitrary shutdowns of offshore development and unrealistic environmental regulations imposed by well-intentioned (but scientifically biased) organizations like the Sierra Club and Friends of the Earth, consumers were sure to suffer. Few of those who waited in lines for gasoline in 1979 probably ever saw the connection between the errors of the past and the shortages of the present. It is simpler to blame either the oil companies, the government, or both. Although it is foolish to blame a whole state for the gullibility of the majority, if a state could be blamed, no state in the Union is more deserving than California of its past energy plight.

Regardless of the damage to the public interest that inevitably followed the defeat of conservation regulation, the damage to those of us who stuck out our necks for the referendum was plain enough. Politics is played in companies and within industries as well as in government, and they were played against those of us who fought for a conservation statute. It hit Ralph Davies like a ton of bricks. His company enemies—led by Standard's new president, whom I will describe at the end of the chapter—sent him to Coventry. He retained his title of vice-president, his office, and his secretary, but his responsibilities were removed. This was a rough time for my associate and dear friend.

Wearing a Different Legal Hat

For me as a lawyer, the referendum defeat meant a stint in the legal salt mines. There were problems aplenty on this level too. Standard had been charged with unfair labor practices at its El Segundo refinery. My job was to defend the company. Warring factions among the labor unions provided an opportunity to settle, but a jurisdictional row developed before the National Labor Board. Hard feelings ran wild between the Oil Workers (an industrial union) and the Pipe Fitters, Boiler Makers, and Carpenters (all craft unions). The unions could agree on only one thing—that the Standard Oil Employees Association was a company union.

All unions—company, craft, and industrial—had some standing with the labor board, but the company itself had none. When I rose to argue for the company, I got no further than to state my name and that of my law firm. The chairman leaned across the bench with the words, "So you represent the Standard Oil Company, eh!" I had no further opportunity to present any argument, so I kept quiet. The deck was stacked against Standard, but I thought it would be better to let the record show the board's bias. The hearing adjourned at the end of several days of wrangling between the unions, and I ran around their flank to talk to the general counsel for the board, who happened to be an old colleague of mine in the government. I relayed to him the state of the record, suggesting I would enjoy taking the case all the way to the Supreme Court. Moreover, if the board decided in favor of any one of the unions, it would incur the anger of all the others. So, why not hold that nobody had asked for an appropriate bargaining unit (which previously had been ruled under similar circumstances involving the Shell Oil Company). The strategy worked: the board dismissed the case, and we went our way rejoicing. Years later, the Oil Workers Union prevailed, but at least I did not have to bear the burden.

One example of how matters were once managed arose out of this early labor case. My immediate client was the manager of the El Segundo refinery, who also happened to be the brother of the vice-president in charge of manufacturing for Standard. When we had concluded the case, some means had to be found to supplement the fee of an attorney who had represented one of the warring unions. My client asked what we were going to do about him. I stated I had retained him to examine some land titles, but in the meantime, I proposed to pay him in advance—in cash. We both agreed to loan the cash into our current expense accounts; but a month later I had a call from my client's brother. I found him in his office with my expense account and that of his brother on his desk. He looked up and asked, "What-in-hell did you and my brother do in Washington—sleep in a gold bed?" I answered, "Dick, do you really want to know? If anyone asks me the right question, as an attorney, I can plead privilege, but you can't." No one had to hit Dick with a brick. He laughed and replied, "Forget it."

After the defeat of the California conservation law, both the firm and I found it best for me to retire to the straightforward practice of law. This meant, however painful, taking off my oil man's hat. It was an interesting *interregnum*.

One of Pillsbury's clients was the Golden State Company, a minority owned by Standard of California as a side investment, thanks to Ralph Davies. Under the direction of Frank Buck, Golden State dominated the California market in not only fresh milk and ice cream but dried milk, canned milk, cheese, and powdered ice cream mix.

Golden State was sued by a competing dairy in Santa Barbara for alleged violations of the state's antitrust laws. It was charged with conspiring with its truck drivers, all members of the teamsters union, to shut down its competitor with a picket line claiming unfair labor practices. This action was accompanied by the drivers' soliciting the customers of the picketed competitor, from which both benefitted. The competitor sought several million dollars' worth of damages, tripled.

In casting about for the best defense, I remembered some Supreme Court cases holding that so long as there was no violence, picketing and charges of unfair practices were a proper exercise of the right of freedom of speech. On these grounds, I demurred to the plaintiff's complaint. In law school we always translated a demurrer to mean—even if all the plaintiff's allegations are true, "so what?" Some of my partners were horrified because this meant we were on the same side as the teamsters union. The simple truth, however, was that we were. After I finished my legal argument, the counsel for the teamsters simply adopted my argument and sat down. The judge said (with a twinkle in his eye) that he found this an amusing case because "I find on the same side of the same case at the same time the Golden State Milk Company, the Teamsters Union and Pillsbury, Madison & Sutro. But be that as it may, the demurrer is sustained."

I was always thankful that I did not have to try the case on its merits, as I was fearful of what those truck drivers and perhaps some of the company managers might have said on the stand. The ruling of the trial court was appealed, but was sustained by the Supreme Court of California several years later, when I had once again donned my "legal-oil" hat. The case helped me with my partners by proving I had some strictly legal abilities outside the oil industry.

Another matter for Golden State involved mixing a minor amount of carotene with whole milk and marketing it as Golden V. Carotene was yellow and a natural source of certain vitamins. Golden V was sold by the pint at double the price of just plain milk. It was highly advertised. Some of our competitors complained to the California regulatory authorities that we had violated an old state law prohibiting the modification of milk. Technically, the complaint was valid, even though the statute was passed primarily to stop watering milk down. With a big volume of Golden V and significant profits on the line, we lost all our legal battles. This left only legislative recourse in Sacramento. I sallied forth and for weeks got almost nowhere with amending the statute. Late one afternoon the senator from Fresno walked into a small group of legislators and with a single off-color joke about Golden V's rejuvenating powers turned the day. When my bill was called up, it passed amid gales of laughter. I never told my client how we won: not so much on the merits as on the off-color story. For ourselves, we did not forget to send the client a different kind of bill— a big one—which was gratefully paid. Even in areas remote from oil, it is good to be good but better to be lucky. The luck of this draw did earn me some brownie points with my partners.

During this period when I'd left (temporarily as it turned out) the complex problems of conservation and competition in oil, I was drawn into a related but different kind of competition—a battle for leases in the strictly gas-producing end of Standard's business. By the late 1930s, no major dry-gas field had ever been found in California, as most of the natural gas production in the state had been associated with oil. Both Standard and Amerada Petroleum sent their geophysical crews into the Sacramento Valley looking for structures likely to trap either oil, gas, or both.

Near the town of Rio Vista a big structure was indicated, though Standard and Amerada interpreted their geophysics differently. Standard was close, but Amerada was on target. (Even today geophysics and its interpretation is anything but an exact science.) Amerada drilled the first exploratory well and stumbled upon the greatest dry gas field in California. Their leases appeared to cover the best of the indicated extent of a field, which later was found to contain trillions of feet of natural gas—just outside Sacramento and well within pipeline range of San Francisco, a major market for natural gas.

Standard had nothing but leases just off the producing formation. But as almost always happens, there were holes in Amerada's block of leases. A few little and some big land owners had held out, and there were acres and acres of unleased state river lands where the Sacramento River wandered in and around the indicated structure. With others, I was assigned the task of trying to rescue the company from its geophysical mistakes and misinterpretation. Fortunately, Amerada was concerned about increasing its bonus and royalty scale, fearing the ire of landowners who had leased their mineral rights under less favorable terms. I had no such worries for Standard. All we had to do was devise new and better lease clauses than those which had been used: profit-sharing deals and high royalties in excess of 50 percent—then an almost unheard of rate until you stopped to remember that these lands appeared to be proven for production and that gas wells flow with little operating expense compared with oil wells which sooner or later must be pumped. As it turned out, we ended up with almost half of the Rio Vista gas field, though not on as favorable terms as Amerada, by virtue of their being right in the first place.

The terms of some of the leases we devised were sufficiently complex as to do credit to a later generation of government regulators. Ours, at least, were not ambiguous. As one of my associates said of one of Standard's landmen: "Howard, you can split a legal hair, but George Schroeder can make a shaving brush out of the two pieces." We were not, however, completely successful in leasing all the holes in Amerada's block. We left 1.16 of an acre fronting on the river where we held a lease from the state. The owner of a cement plant leased this one-acre parcel and drilled a well. He proposed to take 20 million feet a day from this one-acre parcel to supply the total requirements of his plant. This would not only drain gas from everyone around him, but it also threatened to create a pressure sink that would jeopardize the ultimate productivity of the field as a whole.

When it came to wide-open oil or gas production, there were no ten commandments, but only the law of the jungle in California. The situation seemed hopeless, until the owner of this lone well applied to the state lands commission for a pipeline permit to cross state river lands to connect his well to his cement plant. I took on the task of resisting the permit, explaining that the state itself would suffer the greatest damage from drainage, particularly since its lands commanded the highest royalty rate in the field. The commission held a public hearing, and our opponent packed the hearing room with some farmers (and their sisters, cousins, and aunts), who had been promised a little free gas along the proposed pipeline route. As always, their attack was leveled at the wicked monster, Standard Oil. For two days I listened in silence, and, just before the hearing was about to close, I was asked by the head of Standard's gas department why I didn't seem to be worried and why I hadn't said anything. I told him no purpose would be served by building up the opposition and that I knew the permits would be denied by a two to one vote. Just before the end, I expected to present our case for about thirty minutes just to "protect the record" for the two members of the commission who already understood where the state's interest rested. The record was protected, and the vote was two to one for Standard.

Changing of the Guard at Standard

Having lost, our opponent tried to cut the ground out from under my feet by complaining to the new president of Standard, Harry Collier, that I was persecuting him. I was called on the carpet to explain. I took a pencil, sketched a small map of the gas field and pointed to the one-acre parcel offsetting thousands of other productive acres. I remember saying, "You understand the drainage that would occur if the cement plant owner opened his well wide-open on one acre." I was met with a question: "What do you mean by drainage?"

I almost fell through the floor. I actually had to draw an idealized picture of an oil and gas field to explain one of the most elementary principles of reservoir mechanics. Going back to my office, I relayed my shock to my senior partner. Here was the titular head of one of the greatest oil-producing companies in the world who clearly did not understand what "drainage" meant. My partner looked up with the remark, "Did that surprise you?" He was older and wiser than I, and later I too learned that you often become the chief executive officer of a large company simply by living long enough, playing corporate politics, and exercising extreme care never to risk failure or criticism. A week or so later, I successfully persuaded Standard's president that it would make sense if we made a deal to buy out our adversary at Rio Vista for what he'd spent rather than leave him to die on the vine. Once you've won, leave as little bad blood behind you as possible.

In my short time representing Standard of California, three different men had been president of the company. The first was Kenneth R. Kingsbury, a tough second-generation oilman, part of the old Standard oil group only once removed from John D. Rockefeller himself. Kingsbury, who became president of Standard in 1919 at the age of forty-three, knew the business from the ground up. Always immaculately dressed, a white carnation in his buttonhole, he was the picture of the successful industrialist in the first quarter of this century. He

Standard's presidents (*clockwise from upper left:*) Kenneth R. Kingsbury, William Berg, Ralph K. Davies, and Harry Collier.
Courtesy Chevron Corporate Library

graduated from Princeton (class of 1896), and he could put away a half-dozen Gibsons (mixed about twenty to one) and never appear to have had a drink. Ralph Davies and I used to share a few with him, which he mixed himself in a big pitcher at the Pacific Union Club. We would have to walk ourselves up and down those San Francisco hills for an hour or so afterwards, but Ken Kingsbury never showed a sign of even one Gibson, either at cocktail time or the next morning.

Kingsbury's son-in-law, something of an artist, once lectured him about artistic appreciation. Ken turned to him and said, "Young man, the Standard Oil Company is my painting, and I am proud of it." Kingsbury died in his dinner jacket from a heart attack on a pleasure cruise returning from New York to California on November 22, 1937, at the age of sixty-three and just one month short of his fortieth anniversary with the company. Had he lived a few years longer, the succession of Standard certainly would have been different. No one, not even the so-called principal stockholder (John D. Rockefeller, Jr.), ever told Kingsbury what to do.

Shortly after Kingsbury's death, Bill Berg became president. Bill started as a stenographer in the production department in 1902 and made his mark by saving some critical records in a wheelbarrow at the time of the great San Francisco earthquake and fire of 1906. Specializing in production, he became a director in 1924 and vice-president three years later. It was said that when a Standard president died, everyone moved up a notch and a new office boy was hired.

Berg was no Kingsbury. He could never have built a company or painted the Standard picture as Kingsbury did. On the other hand, you would not have had to explain what "drainage" meant. He knew production, but marketing and politics always scared him to death. He was decent, honest, and hard-working but without great imagination— he was more a custodian of great assets than a builder or risk taker. The job may have been too big for him. He too died of a heart attack, on June 27, 1940, at age fifty-eight, less than three years into his presidency.

After Berg, who was next? It was generally expected to be Ralph K. Davies, one of four vice-presidents and the protégé of Kingsbury himself. Based on the record of accomplishments, there would have been no question. Ralph joined Standard in Fresno at age fifteen as a clerk, finishing high school by night. He was noticed early and brought to the home office in San Francisco in 1917 as a special assistant in sales management. He became the youngest director in Standard's history at age thirty-three. In July, 1935, he became vice-president of sales, a key position. As Ralph put it, in an integrated company it is all bookkeeping before the final sale.

To be prominently mentioned as the next president of any substantial company is often the "kiss of death." It is like running for high political office too soon. All the others who would like to be president unite to try to stop the front-runner, and so it happened to Ralph Davies. The loss of the conservation statute, as I mentioned before, was used by some of his opponents. Even his financial brilliance did not make friends and influence people. As one of his subordinates once said, "Why is it when I present him columns of figures, he always picks the one figure I don't want to talk about." Actually, Ralph was elected president by a majority of the board soon after Berg's death. But a few of the minority, no doubt fearing what might happen to themselves under a strong chief executive, united to lobby New York to reverse the decision. Even in the late 1930s there were those who still asked in a whisper, "What are the wishes of the principal stockholder?" When those wishes were made known, the board (on July 8, 1940) reversed itself to elect Harry D. Collier, who was, theoretically, a year away from mandatory retirement at the age of sixty-four.

Collier, who began with Standard in 1903 and became a vice-president in 1931, came from a bookkeeping background. Kingsbury once told Davies in my presence that he often had to tell him to go back and sit on his stool. A corporate fool, however, Harry was not. He promptly relieved his rival, Davies, of all duties and put him out to pasture. He promoted his own friends and cut down or retired those who did not pledge allegiance. Even some of Ralph's supporters changed sides. Even Al Weil, our friendly competitor, with whom we had fought so many battles together, failed us. He told me a few years later that New York had called him about Davies before the second vote. Al confessed to what he called a terrible mistake. "I told the board to let Harry Collier have the presidency for the year before he had to retire." Al added, "It never occurred to me that the old boy would suspend the retirement rule for himself to stay on."

Collier did—until he was in his seventies. He was president until 1945, chairman of the board through 1949, and chairman of the finance committee until 1951. He resigned from the board in 1956 at age eighty and kept his Standard office until his death on January 30, 1959.

STANDARD OF CALIFORNIA

Personally, I have always wondered whether the New York powers really wanted a strong chief executive. I never observed any eastern influence in day-to-day competitive operations, but, at that time, I doubt that any chief executive could be elected without the approval of those eastern powers. When Standard of California (under Collier) divided up its position in Saudi Arabia with Jersey and Mobil, it seemed to me that somebody was saying, "Now boys, fun is fun but just keep us together in the world market." The invisible powers wanted all of the action in the Standard fold—or so it appeared to me.

Kingsbury, who never took orders, probably rolled over in his grave over this buy-in. After all, it was Kingsbury who earned California its place in the Middle Eastern sun. It was Kingsbury who had brought Shell and Jersey to heel in British Columbia to teach them to reckon with a new competitive force in that same Middle East. When Aramco was formed—out of Jersey, Mobil, California and the Texas Company—to own and operate all the concessions in Saudi Arabia previously owned first by Standard of California and later by California and the Texas Company, Davies and I looked at each other. Could we read between the lines? After all, Kingsbury and Davies had taken in the Texas Company because they had the markets and we had the production they needed. Neither of us saw any reason why California and Texas needed Mobil and Jersey. By then, however, neither Ralph nor I was close enough to the throne to solve that mystery.

The nearest Ralph (and, to a lesser extent, I) ever got to a throne room occurred shortly after the board of Standard of California reversed its decision and elected Collier president. Ralph asked for and got an appointment with John D. Rockefeller, Jr. I was in New York with Ralph, and when he got back to the hotel, I asked what he had learned at the meeting. "Nothing," he said. "I was treated with the greatest courtesy. At the end, John D. asked to see me to the elevator." I made almost verbatim notes of the conversations with Ralph as he relayed them a few minutes after the meeting. Ralph and I adopted an expression derived from that meeting. When some troublesome character descended on either of us, we would ask, "Did you see him to the elevator?"

The Collier era was surely different than what the Davies era would have been. Ralph's special assistant, William Bates, later wrote why he left the company soon after Ralph was pushed aside: "The opportunity for imagination, the very idea of inspiration, the forthright atmosphere and challenge of free thinking had evaporated."

Ralph had narrowly missed the top position of one of the leading oil companies in the world. By the same margin I had missed being its number-two man. Like Ken Kingsbury and Oscar Sutro before us, Davies would have been president and I his principal vice-president and general counsel. But other things were in store for both of us.

CHAPTER FOUR

Back With Government: War Petroleum Planning, 1941–44

After Harry Collier became president of Standard of California in July, 1940, I watched my good friend Ralph Davies suffer "in Coventry" for months. He drew his salary, but nothing more. Take away a person's life work, and money becomes unimportant. Even Ralph's friends in the company—and there were many—did not dare be seen with him for fear they would land in Collier's doghouse. I went back to practicing law, but Ralph had no choice except to quit—but quitting was not part of his make-up. At least he and I could still have lunch together, even in public, without fearing the consequences. As a partner of a leading San Francisco law firm, I refused to worry about anything but my clients and cases. I was removed, at least for the moment, from the mainstream of internal corporate political maneuvering.

Planning for War Planning

One day at lunch at the Stock Exchange Club in San Francisco, after the European War had broken out, I remarked to Ralph that it seemed inevitable that the United States would become involved. We both asked the same question in almost the same breath—if war came, who would run the petroleum industry? An idea struck: why not start figuring out how it should be done and try to position ourselves to do it? I had the ideal governmental background—I knew the bureaucracy like an open book. We both knew the oil industry. We took it from there. Night after night, we labored to propose an organization within the government which would take charge in the event of war.

Our homework was done on a yellow pad over dinner tables in a small, out-of-the-way restaurant in San Francisco called The Manger. Perhaps the name was prophetic, as the restaurant provided the birthplace for what later became the Petroleum Administration for War and the Petroleum Industry War Council.

As a lawyer, the legal problem of establishing a planning apparatus held the same fascination as the "extralegal" devices I had helped to invent. How could some kind of a newly created Washington agency, without any specific statutory authority, lay the groundwork for directing the entire petroleum industry to help fight a war we knew was coming? We had some good examples—the policy for selling destroyers to Great Britain was one; "Lend-Lease" was another. The outlines of future controls applicable to all critical materials were already emerging. Here was the key: if the president would direct the secretary of the Interior to establish an office to coordinate all governmental activities in any way related to oil, and ask the same office to coordinate the activities of the industry itself, we could begin getting ready for what was coming. The magic word was "coordinate," as opposed to "regulate" or "order"—words with too definite a legal meaning and requiring specific statutory authority.

I still have in my own handwriting the draft of a letter from President Roosevelt to Interior Secretary Harold Ickes directing Ickes to establish the Office of the Petroleum Coordinator to coordinate all governmental activities relating to petroleum supplies and recommend to the petroleum industry such actions it should undertake to assure such supplies.

Even before our entry into World War II, Roosevelt proclaimed that the United States should become the arsenal of democracy. Coordination, rather than blind competition, was needed to produce, refine, transport and distribute the enormous quantities needed for the war effort. We had surpluses, but not enough and not of the right kinds in the right places at the right time. Moreover, the petroleum industry itself needed the tools—both material and financial—to produce what was needed to both fight the war on the battlefront and maintain a domestic infrastructure upon which the production of everything from food to ammunition depended. Nothing less than a full-fledged partnership between all of government and all of the oil industry gave any promise of making ends meet. Given federal control of supplies from the oil industry and critical materials supplied to that industry—steel, pipe, pumps, chemicals, and equipment for example—it could be done. The letter from the president to Ickes set forth in detail how the Secretary should do it and told all other government agencies to cooperate with the secretary on all matters relating to the oil industry. The letter laid out in polite but explicit terms (particularly to these other agencies of government) that oil was henceforth to be the secretary's bailiwick. No statutory authority was needed to coordinate the whole industry, given our control (direct or indirect) of the critical materials without which firms could not long survive. Perhaps it was extralegal, but it was practical. It may have driven the lawyers crazy, but in the final analysis it helped to win the war.

Having satisfied ourselves how a centralized federal agency might effectively fight the wartime battles of petroleum, how could we go about getting the proposed letter from the president into his hands? No major companies would be likely to help us. One potential ally, Dr. Robert Wilson (of Standard of Indiana), had already been selected to head the oil division of the soon-to-be all-powerful War Production Board. Naturally, the eastern establishments of the large oil companies would want no up-starts from California suggesting what they should do. Still less would they welcome power in the hands of an old firebrand like Ickes, who had sponsored federal control of the industry during the NRA, and whom the industry regarded as having let them down in the Madison trial.

Cut off from our major friends, we turned to an independent, who was none other than our old political ally, Ed Pauley. His Democratic connections were impeccable—he was a money-raiser for the party who had labored long and effectively in party circles from the lowest to the highest levels. A personal friend of both President Roosevelt and the new vice-president, Harry Truman (whom Pauley had helped to nominate in Chicago), Pauley had easy access to the White House. He used it. Like a bolt from the blue, the president's letter to Ickes was issued on May 28, 1941, one day after FDR's declaration of an "unlimited national emergency." No "informed" leaks preceded it, and for once the Eastern majors were caught off guard.

Ed easily persuaded Roosevelt to make Ickes the Petroleum Coordinator. Ickes was delighted. He had partly fallen out of the President's good graces and welcomed the chance to mend his fences. Pauley was quick to persuade Ickes that his friend, Ralph Davies, was the ideal man to serve as the secretary's deputy to, in effect, run the whole show. Pauley probably told Ickes that Ralph, like the secretary, had found himself in a dog-house and certainly would not serve as anybody's stooge if he entered government service. Davies qualified as an oil industry independent—a big advantage politically—despite formally working for a major.

For once the deep laid plans of a few mice and men worked right on schedule. Ickes became Petroleum Coordinator on May 28, 1941. Davies, taking a leave of absence from California Standard on June 6, 1941, became his deputy on June 10. For Davies it was out from Coventry and into a brave new world with, potentially, the whole nation's oil industry to coordinate. A week later, the two of them were on the phone asking me to take a leave of absence from my law firm for a few months to "help them get the new agency started" as its chief counsel.

The New Chief Counsel

I got the leave from Pillsbury, Madison and Sutro, but it was granted without pay. An obscure statute prohibited any federal employees from receiving any income from the prosecution of any claim against the United States. Since my firm often prosecuted tax cases against the United States and sometimes won, my partners decided that I could not draw any monies from my firm as long as I was employed by the government. I had to be "employed" by the government; Ickes would take no "dollar a year" men on the theory you could not serve two masters. Even though I disagreed with my partners' interpretation of an old, obscure statute obviously aimed at stopping bribery, I abided by their decision.

On July 25, 1941, my income dropped from well over a hundred thousand dollars to eighty-six hundred a year. Fortunately, I had a few securities good for something less than two hundred dollars a month. I sent my government checks to the family and lived on the two hundred

Harold Ickes. *Courtesy Harvard University Press*

and quickly learned, again, how the street cars and buses ran. It was really less of a hardship than it seemed at the time. We were all working sixteen hours a day, seven days a week. There was no time to spend any money anyway. When one of our associates left during the war because he could not balance his budget on a government salary, I laughingly told him there were two ways to balance a budget—earn more or spend less. I asked him if he ever tried the second way, since I found from experience that it worked.

I would be less than honest if I didn't confess to a small diversion to help escape from the rigors of an austere budget. Everette De-Golyer, wealthy in his own right, founded what became known as the Eat on Industry (EOI) Club. As De put it, "You don't have to be very important in this town to get a free lunch. But when you become so important that on the same day you get a free breakfast, lunch, and dinner with different members of the industry, then and only then can you be admitted to the EOI Club." When one of our members left to return to his company, we would present him with a certificate transferring his membership from the status of an active member to a "sustaining" member. Once, when one of these members, from Mobil, came back to Washington, a group of us insisted on a sumptuous repast at the Mayflower. We ran up a big bill. Each of the "guests" autographed the bill, and we sent it to Brewster Jennings, head of Mobil, who had been a member of the Club when he served with us in Washington, with the words: "Dear Brewster, just so you will know we have not lost our touch."

Arriving back in Washington in the early summer of 1941 was, in some respects, like returning home. There was the same secretary, the same old Department of Interior (now ensconced in a new building), the same old antitrust division in the Department of Justice still headed by my old Yale colleague, Thurman Arnold, the same old humid, hot Washington climate, and the same old battles between different departments, bureaus, commissions and agencies. It was said during the war that the Germans never bombed Washington because it would increase efficiency if some of the dozens of overlapping agencies were destroyed.

The new Office of Petroleum for National Defense Coordination (OPC in the Washington vernacular) was housed in the new Interior building. Except for the deputy administrator, who inherited the assistant secretary's corner office, the rest of us started with bare walls. I quickly got a makeshift desk, a chair, and those absolutely essential government tools—a telephone and a typewriter. Nothing more appeared for months, but it hardly mattered. I've done some of my best work in the worst offices and some of the worst in elegant offices.

Early in the game, Ralph Davies asked me how to deal with the conflicts and confusions of overlapping agencies, as each tried to assert jurisdiction over some phase of the oil business. Even after the Department of Energy was created in October, 1977, to consolidate the agency hodgepodge, the problem remained. Not even Pres. Jimmy Carter, despite all his protestations to the contrary, had the strength to keep his many children from fighting each other rather than solving problems. For example, the Federal Energy Regulatory Commission, which regulated natural gas and electricity, was set up as an independent agency. Under wartime pressures, however, we did finally get the infighting laid to rest—not by compromising but by usurping and concentrating at one single point almost all the powers of the government concerning oil. "The trick in Washington," I told Ralph, "is not to ask for authority—just grab the ball and run like hell with it before these other fellows know you've got it." It worked, but it took an able administrator like Davies, who knew his industry, a cabinet officer like Ickes, whom all Washington feared, and a president like Roosevelt, who was smart enough to let Ickes and Davies run with the ball.

When we started, hardly a day passed without Ickes asking us whether we had run poor Dr. Wilson out of town yet. "Doc" Wilson, whom I later got to know and respect from his position at the War Production Board, was ostensibly in charge of critical materials for the oil industry. He probably never figured out how the Petroleum Coordinator absorbed all his functions. When his job became merely ministerial—meaning he was expected to allocate materials as the Office of the Petroleum Coordinator "recommended"—he went back to Standard of Indiana, and Secretary Ickes could say, "Well we don't have to worry about him anymore."

Two matters commanded our immediate attention in the formative period of the OPC—choosing a government staff and an industry staff to advise us. Both groups had to be highly qualified—not in the present-day sense of knowing nothing about the industry for fear they might be prejudiced but just the reverse. They should be selected solely on the basis of how much they knew about and what experience they had in the industry. On the government side, we drafted from the industry not those with important-sounding titles but independents, top middle management, and technical talent from all over the nation. Some of us had ranged far and wide over the oil industry both at home and abroad, and we knew enough about its members to start choosing a staff from both government and industry ranks. Good selections helped us find others—poor selections usually eliminated themselves when they found their task too demanding. Some years later, I could honestly claim that with the staff we assembled I could run most major companies better than most of them were run.

On the industry side, we tore a leaf from the past. As Justice Oliver Wendell Holmes once put it, "a page of experience is worth a volume of logic." So we established an industry advisory group, the Petroleum Industry Council for National Defense, which was composed of the top executives of oil companies—independent and major, integrated and non-integrated—from every branch of the business and from all over the country. We added the best individual entrepreneurs and trade association leaders. When that council met, as it did regularly throughout the war, there was more oil horsepower assembled in a single room than I have ever seen before or since. No member was permitted a proxy or a substitute. Invariably they all came in person. Perhaps some were afraid not to attend.

To match the governmental and industry organization in Washington, we divided the country into five districts along natural industry lines—the East Coast, the Middle West, the Gulf Coast, the Rocky Mountains, and the West Coast. The OPC established a regional office in each of the five districts, which were divided, as in our Washington office, into divisions to deal with production, refining, transportation and marketing. All were staffed with talent drafted from the industry. On the industry side, regional industry committees were appointed to match those in Washington that were headed up by the national advisory council.

Both in Washington and in the field, the government staff and the industry committees worked in tandem. There was almost none of the adversarial attitude so typical of three decades later, when oil and natural gas went through an "energy crisis." Even though it may never be duplicated, it should not be forgotten there was, for a few short years, "a place called Camelot where government and industry did pull together in a partnership."

As the summer of 1941 faded into fall, supply problems—particularly with gasoline—began to intensify. The European war began to drain tankers transporting aviation gasoline and fuel oils, and gasoline proved to be the most sensitive political nerve in the whole assemblage of petroleum products. As tankers were diverted to the war zones and frequently sunk, the East Coast of the United States, largely dependent on tankers to move product from Gulf Coast refineries to points of consumption, began to suffer spot shortages of gasoline, prompting the usual cries of "phony shortages" from the public. Once when my deputy had to testify before a Senate Committee for three days running, I went some sixty hours without going to bed. As his lawyer, I had to spend all day with him at the hearing and all night preparing the testimony for the next day. Before the hearing, he had never even been in a courtroom. After the hearing, he came out a veteran. I went to bed and slept the clock around.

It was during this period that we invented a previously unheard of legal document called a "recommendation." It was based on nothing more substantial than the use of that word by the president in the letter establishing the OPC. Since I had used the word in that same letter, I had no compunction about using the word for our purposes. Our recommendation was addressed to the petroleum industry, and it explicitly, specifically stated what we wanted them to do. We dolled up our recommendations and put them in the *Federal Register* in the same way that regulations with the force of law behind them would have been. We cited, without reciting, the president's letter as our authority. I often wondered how my old adversary Big Fish would have ever figured this one out. A lot of good industry lawyers wondered at the problem too.

There was, however, at least one good legal reason for formalizing our recommendations. Having gone through the Madison Oil Case, in which the defendants failed to prove that they had acted in response to a governmental request rather than a "conspiracy," I wanted to leave a clear paper trail demonstrating that, if a group of companies did the same thing at the same time, their actions did not necessarily arise out of a private conspiracy. With the East Coast gasoline shortages, one of our recommendations related to service station closing hours—a 7 P.M.–7 A.M. curfew, which helped a little. We never thought of recommending sales on certain days for odd and even license plate holders, as was implemented by various states in 1974. If we had thought of it, we probably would have tried it. At that time we weren't aware that general gasoline rationing plans hardly ever work for long—not even in wartime, let alone in peacetime. Only if service stations run out of gasoline does rationing ever seem to work.

As summer wore on and fall fast approached, problems in the petroleum industry grew even more hectic. For a time I lived day-by-day at the old Shoreham Hotel, and, when I persuaded DeGolyer to join us as an assistant deputy in charge of exploration—there was no one in the world better qualified—he and I became roommates there. Shortly thereafter, we found a three-room apartment within a block of the Interior Building. It cost all of $77.50 a month, but it was air-conditioned. When I had it furnished for about nine-hundred dollars, I sent De a bill for his half. When he sent me his check, I found a cryptic note; "Howard, does this include a rug?" He had just finished building a half-million–dollar mansion in Dallas, which he referred to as the "Casa DeGolyer" in honor of his fluency in Spanish.

Once we got wind of prospective coffee rationing. We promptly raided the chain stores to stock up some thirty-eight pounds in vacuum tins, knowing that we could not fight a war without coffee. Later when coffee was rationed, the ration application demanded to know how much coffee you had on hand. We solved that one. I would solemnly announce before I signed my form—"DeGolyer, I hereby assign to you all of my right, title, and interest in the coffee hoard." Then I would fill in the blank for coffee on hand with the word "none." De would then repeat the performance to me to claim "none." When the war ended, we still had a few pounds left.

During the fall of 1941, we were laying the basis for most of the activities which would occupy us when the "hot" war really broke out. Material controls, prospective aviation gasoline supplies, overland transportation of oil and its products, tanker controls, crude oil regulation—these were some of the matters with which our "recommendations" dealt. Then suddenly one Sunday afternoon, our signal drills were over. As I walked down the long corridors leading to the deputy's office on December 7, 1941, I was met by Chandler Ide, who told me Japan had bombed Pearl Harbor. My first reaction was to tell Chandler to quit kidding, I was busy. But the radio quickly confirmed his report. Actually, I think I breathed a sigh of relief. Now, I realized, the prologue was over. The real show was about to begin.

BACK WITH GOVERNMENT
Real Wartime Planning Begins

Wartime planning began the next morning, with a meeting of the Petroleum Industry Council for National Defense which had already been set. Following the U.S. declaration of war, the new names for both our organization and our industry advisory group became, respectively, the Office of Petroleum Coordinator for War and the Petroleum Industry War Council. A year later, the OPC was renamed the Petroleum Administration for War (or PAW) and given expanded authority.

Gone were my own hopes for staying at the OPC only six months to help get the show started—clearly there was nothing to do but stay in Washington and help fight the war. Tough as it was, I could later say that it was easier than being stared down by an enemy's machine-gun sights. Personally, I doubt I could have been better trained for the task if someone at Yale had devised the right courses to take. I could claim theoretical and academic training of a high order, government experience as a lawyer, administrative dealings with the oil industry in a time of crisis, and, most recently, detailed, practical contact with almost every phase of this integrated business, both as a lawyer and an oil executive, with one of the largest and best of the integrated, worldwide majors. Likewise, my client and deputy administrator came to his job—young, strong, tough, intelligent, and with an oil background, which started as an office boy with Standard of California at the age of fifteen and ended as one of the top executives of his company while in his early thirties. His misfortune—to find himself in Standard's political doghouse—was the nation's good fortune. Later, when it was all over, even his worst enemies would say no one better could have been found for the job.

Almost immediately we began to lose tankers by the dozens to German submarines along our East Coast. So frightful were the losses and so long the many months before they could be replaced that we had no choice except to order our tankers into port until they could be protected. This action left the coast, and particularly New England, short of essential heating oils and kerosene in the middle of the winter. We prepared to throw the problem at the Petroleum Industry War Council. Riding downtown with Bob Minkler, drafted from General Petroleum to run supply and distribution for the PAW, I noticed him making notes on the back of an envelope, à la Lincoln's Gettysburg Address. After the Council talked itself to death without coming up with a workable solution, Bob rose to read from his notes on the envelope.

He laid out a simple program. He asked the industry to stop all overland backhauls from east to west, move everything from southwest to northeast, whether by barge, truck, tank car or pipeline, and absorb the excess costs. He indicated we would try to find a method of reimbursing the companies. I remember R. H. Colley, the chief executive of the Atlantic Refining Company, frantically figuring on a pad in front of him. With a tremor in his voice, he asked Bob if he had any idea what this program would cost. Bob said he did. Colley indicated his figures showed the cost to be about a half billion dollars annually. Minkler answered, "I have been in Washington only a few months, but back here we refer to that amount as point five." Afterwards we told Bob that, on this scale, a millionaire was worth point zero one. DeGolyer, in turn, nicknamed his friend "Cod Fish Colley" because, like a banker, he had a fishy eye.

It took the Defense Supplies Corporation, run by the Secretary of Commerce Jesse Jones (an old banker himself), a year to work out the reimbursement. To help us develop a formula for measuring excess costs, the Council enlisted the aid of Dr. Alexander Saks. Alex had at least two distinctions. First, he had introduced Albert Einstein to Roosevelt to explain the necessity of developing an atomic bomb before the enemy beat us to the punch. Second, he could use big words—even if he had to invent them—to describe simple ideas more than any man alive. In spite of, though certainly not because of, his wordy presentations, we won out with Jesse Jones. If Uncle Jesse understood Saks, he gave no sign of it.

One other example of ingenuity, coupled with industry cooperation, helped us through a severe New England winter. The slum areas of Boston relied on kerosene for heating and cooking—there was plenty in Texas and almost none in Boston, with no apparent way of getting it there. No way, at least, until J. R. Parten, head of the Transportation Division of the PAW, suggested one to Secretary Ickes on their way to the White House. Why not use steel drums filled with kerosene and loaded into box cars in trainload lots? All we needed to do was to persuade the military to load the drums (they, too, liked to hoard) and convince the oil industry to both fill and drain the drums at daily rates exceeding several tens of thousands of barrels. It was done to the tune of fifty thousand barrels a day and more, and we survived yet another crisis.

If we shift our focus from the East Coast to the West Coast and move upstream in the industry nexus, we can see how California's production practices again reared their ugly heads. This time, however, we had the power to deal with them. "Dealing with them" took the form of insuring that scarce critical materials—pipe, steel, pumps, and the like—would not be wasted in town-lot drilling or other competitive excesses. Back I went to the Central Committee of California Oil Producers, and we cleaned up the allocation formula and established maximum efficient rates of production for the individual wells. Then we denied maintenance and operating supplies to those who failed to abide by the rules set by PAW (with the assistance and cooperation of the Central Committee).

Such problems were not confined to California, however. Illinois, with wholly inadequate statutes, was worse. Even states with conservation laws, such as Texas and Louisiana, posed similar problems. One example concerned the Katy gas condensate field in Texas, which covered three counties and held reserves of several trillion cubic feet. George Hill, president of Houston Oil and a close personal friend of DeGolyer, demanded enough pipe to drill up his small tract on forty-acre spacing. Under Texas law he had a right to do so, regardless of the fact that 1 well per 160 acres could drain the field. We refused to allocate the pipe. I had De sit with me when George came storming into my office screaming about his constitutional rights. He threatened, as he always did, "to meet us at the CO-OUT house." I asked him if he had figured out which CO-OUT house he could get us into, since I was not denying him anything—just considering whether his request would waste material needed in the war effort. I added that his request might take quite a while to consider. De laughed, and George stalked out, white with rage, but he finally came around and unitized his tract with others to conform to 160-acre spacing.

Sometime afterward, when the field had expanded far beyond the original area thought productive, George found himself a large leaseholder rather than a small one. A major competitor wanted material to build a natural gasoline extraction plant for its own account. Since only one plant appeared justified, that company tried to force George to put his gas through its plant on a nominal royalty basis. Now the shoe was on the other foot. We forced George's competitor to take him into their plant in proportion to the gas to be supplied. Overnight De and I became heroes instead of bums. All we did was call the shots the same way, regardless of the parties involved. The contending partners were hardly small fries—the Houston Oil Company, Standard of New Jersey (then Humble Oil in Texas), and Standard of Indiana through its subsidiary, Stanolind.

In Illinois, accounting for 5 percent of national crude output, things were as bad or worse than in California. There were almost no spacing rules and unlimited production—the usual mess. It was said that the McKloskey Lime in the Illinois Basin was only a short-lived, flash-producing horizon. But whoever gave it a chance to be long-lived? As soon as the operators found it productive, they rushed in to drill on twenty-acre spacing, cocked the wells back, sucked in the water, and blew the associated gas into the air. Most other zones were treated in a similar fashion. Here we fell upon another lucky accident, which taught us an administrative trick that we used over and over again. We promulgated an order denying material for any well drilled on closer than forty-acre spacing, unless we granted an exception. From this order we distilled the device of writing a general rule that would fit most cases, and then administering the rule by the exception route. After De and I put the spacing rule into effect, we ran into Charlie Roeser, a prominent and wealthy independent producer, in the lobby of the Shoreham Hotel. He started in on the two of us: "What bunch of damn fools who never ever saw an oil well wrote this crazy forty-acre spacing rule?" De looked at him and said, "I did Charlie, I did." For once in his life, Charlie was left speechless.

Ruling by the exception route was another illustration of what we had learned in the hot oil days—get your hands on the problem at its throat to minimize the number of individual instances requiring day-to-day administrative control. This rule applied to spacing, production practices, marketing facilities, refiner operations, transportation requirements, and maintenance and operating supplies. Write a sound practical general rule, let that apply automatically over a wide area, recognize that no general rule ever works in every practical case, and then deal individually with the exceptions. The exceptional cases gave us enough problems without looking for more or trying to build an enormous bureaucratic staff to deal with a multitude of minutiae.

When a tough exception came along, it was always a "rush" job. Initially such cases carried a tab labeled "rush." As with money during inflation, our "currency" became debased with too many rush tabs. Something with higher priority came to be labeled "urgent." Finally, someone invented a new slip called "frantic." Over at the War Production Board, they had a similar problem with "priority rating." Soon, merely holding a priority rating was not enough. Then came priority ratings A, double-A, and triple-A. Then they started over again. The system was called the "Controlled Materials Plan"—CMP for short. These initials were soon interpreted as "Christ More Paper." Too bad the later Washington energy regulators did not absorb a little administrative history!

One possible weakness in the "exception" approach kept me awake at night. Often an approved exception could be worth millions of dollars to the applicant. I was always fearful that someone on our staff would err or, even worse, be reached by the industry in some inappropriate way. We set up a system under which every approved exception had to pass through my office to be checked for legal form before it was issued. In fact, "legal form" was the last thing that concerned us. An exception was only valid if it bore the signature of the deputy administrator. Only a trusted priority specialist, knowledgeable in the business—none other than the former executive secretary of the Central Committee of California Oil Producers and an oilwoman I describe in the final chapter, Bettye Bohanon—had a facsimile of the deputy's signature under lock and key. She checked each exception first for anything questionable. If she had questions, she raised them with Justin Wolf, my top assistant chief counsel, and, if he had any, he brought it to me. To the best of my knowledge, nothing improper ever got through this screen. If something had, I'm sure the media would have treated us no better during World War II than they do today when some minor suspicion attaches to any oil-related official who commits the slightest indiscretion.

No one should labor under the illusion that the PAW administered its duties solely by direct order. On the contrary, we had the people who knew the people who ran the industry. The box car movements of kerosene to New England were initiated without the scratch of a legal pen. So, too, were the overland movements of crude and products at the time of the submarine blockade of the East Coast. Once, Sen. Ken McKellar, a Democrat from Tennessee, demanded that I locate a few thousand barrels a day of Illinois Basin crude oil for a small refiner in his state. I knew what he didn't know I knew—that he had a personal interest in the refiner. One company from whom it could be secured was Standard of Ohio, and I called my old friend, Sydney Swensrud, Standard's chairman, to ask for it. Sydney, a notoriously tough executive, told me in no uncertain terms I had no right to ask for the crude. He was right, and I said so—but I said I wanted it anyway. I got it, and, after the war, I told him why I had needed it. He then understood. Wars do not abolish politics, and McKellar was too powerful in the Senate and too important to the PAW for any of us to ignore him.

The Big Inch and Little Big Inch Pipelines

The building of the "Big Inch" pipeline was another wartime victory to come out of the industry-government partnership in the early stages of the conflict. Tank cars, barges, and trucks were only temporary expedients to survive the first winter of the war. Feeding the military in Europe by these methods, however, was about as effective as using an eye dropper. Several major petroleum pipelines were needed to flow large volumes to the Eastern seaboard—the central domestic consumption area and staging point for the European theater of war. The idea was thus born to build a 1,400 mile "Big Inch" crude oil line (24 inches in diameter—the largest such line of its day) from Longview, Texas, to Norris City, Illinois. By fanning out supply to the East Coast, tanker hauls to critical East Coast refineries could be displaced. The national security need was obvious: German planes could not disable our pipelines like their submarines could our tankers.

Negotiations to build the line were headed by J. R. Parten for the PAW and W. Alton (Pete) Jones, head of Cities Service, for an industry consortium. The original plan was for the industry consortium to build and own the Big Inch with financing from the Defense Plant Corporation or Defense Supplies Corporation, both headed by Jesse Jones. Jesse was much too shrewd to put up all the money and not have the government own at least some of it. He and Davies debated about how much. Davies, in one of his classic short statements, said, "Mr. Secretary, the best formula I know is fifty-fifty—it always sounds fair whether it is or is not." So, for the moment at least, it was decided. But Pete failed to sell all of his prospective industry partners to go

The Big Inch pipeline was built in all kinds of weather and traversed a variety of terrain. *Picture from John W. Frey and H. Chandler Ide,* A History of the Petroleum Administration for War *(Washington, D.C.: GPO, 1946).*

along. Everyone was short of steel and pipe. Some, building or owning oil tankers, feared the competition of a large diameter pipeline after the war. As a result, it fell back on the government through one of Jesse Jones's arms—the Defense Plant Corporation—to build and own what became the Big Inch and Little Big Inch lines. This left the oil companies in the role of a managing contractor. The work began in August, 1942. They laid the pipe over mountains, flood plains, and rivers—much of it in the dead of winter. The world's largest oil pipeline, designed for 300,000 barrels per day, began deliveries on August 14, 1943. The total project cost was seventy-nine million dollars.

Even after the die was cast to start the Big Inch, the tanker interests almost blocked the project at the War Production Board, the wartime agency responsible for allocating steel and pipe. A compromise allocated the critical materials to build as far as Norris City, Illinois, a major railroad junction point about halfway from Texas to Philadelphia and New York. From here we could temporarily use tank cars to ship the crude in trainload lots the rest of the way to the East Coast. This was not good, but it was better than nothing. In the PAW we figured, if we got a line this far, no power on earth could stop us from going the rest of the way.

One bit of Washington infighting exemplifies how the final card was played. Donald Nelson, then Chairman of the War Production Board, was the key. Secretary Ickes played the card like a master. He leaked to Drew Pearson, a feared political columnist (and the Ralph Nader of his day), that perhaps Nelson had become the "tool" of certain major oil companies opposed to the final leg of the project. Pearson even named names.

I knew Nelson, and one day I got a call from his assistant. Would I come see Mr. Nelson on a confidential basis? He wanted to know whether I could get my secretary to stop calling him names in the newspaper, if he allocated the steel to finish the Big Inch. Of course I visited, and after my necessary denials of any knowledge of any such foul deed, I told him that, if he promised us the steel, I would talk to my secretary about persuading the guilty parties to cease and desist. He promised the steel. With a grin, I asked Ickes for his best efforts to get the media to lay off Nelson. I added: "We may need this card again."

We did replay the card a couple of times—but only on critical matters like the 100-octane aviation gasoline program, without which the war could not have been won. My technique was simple. After clearing the matter with Ickes, I would call Nelson confidentially, tell him about our problems with his board and end with words to the effect that I did not know how much longer I could keep my secretary out of the newspaper. That trump always did the trick. The trick for me was never to use this ace in the hole except when we were desperate. It is always tempting, even in a bridge game, to lay down the cards too soon or too often.

Later came the "Little Big Inch" (a mere 20 inches in diameter) to move the most valuable petroleum products—aviation gasolines, motor fuels and light heating oils—from the Southwest to the East

Coast. It was constructed between April, 1943, and March, 1944, at a daily design capacity of 225,000 barrels. Industry opposition again arose for competitive reasons. Those with an interest in the Great Lakes Pipeline System, a refined products pipeline from the Gulf Coast to the heart of the Midwest, insisted that if a Little Big Inch were built, it should first start from the East Coast and run to Norris City, Illinois, where it could be filled by tank cars from the Gulf.

This seemed silly. As in the case of the Big Inch, if tank cars were to be used for part of the movement, it was obviously better to fan out refined products from a middle western point over better railroads to many Eastern destinations than to try to use second class railroads to move products northward from the Gulf to that same middle western destination. Finally, I figured out the real reason for our opposition's argument.

I got myself invited to dinner with Barney Majewski, the head of Deep Rock Oil, who was one of those loudly arguing for the first leg to be built from New York to Norris City. For whatever reason, my good friend Barney was lobbying on behalf of Phillips Petroleum Company. I asked him not to testify before a congressional committee in favor of such a foolish proposal. I had to lay it on the line. If he did, I promised to follow him on the stand. I would start by saying I hated to air dirty linen in public, but this whole bit of nonsense originated with Phillips. In the event the war ended before the second leg of the Little Big Inch was built, they feared a potential competitor to the Great Lakes product pipeline, a Phillips affiliate, which moved petroleum products from the central Midwest northward. Barney was proud of his independence, and he didn't want to be suspected of being a stooge for "Uncle Frank" Phillips (the Chairman of Phillips).

Barney laid off, but Phillips did not give up so easily. A day or so later I was invited to dinner by "Uncle Frank" and K. S. "Boots" Adams, his president. Boots tried to lay it on the line to me. If I continued to oppose building from New York westward, he would "see that I was run out of Washington." I asked if he could guarantee that because I had been trying to get out of Washington for months. We all laughed, had a good dinner, and that was the end of the whole matter.

The Little Big Inch was first built from the Gulf to Illinois and, later, to the East Coast. Strangely enough, this facility, combined with another leg built to Chicago, ended up, in peacetime, in private hands competing with other carriers and other companies for markets and product movements in both the Midwest and the East Coast. Had our Midwest opposition foreseen this result, they might have felt better, though hardly pleased. As with so many battles, this one broke up no beautiful friendships. We all laughed about it after the war was won, and we were all back cutting each other's throats in the marketplace.

Here, once again, I met myself "coming back from the other direction." Like with the Kettleman Hills episode in California, when I left the government I thought I had left behind the competitive problems raised by the Big and Little Big Inch pipelines, but I was wrong. As president of Ashland, a midwestern refiner-marketer, I found myself face-to-face with the competitive aspects of the postwar uses of these lines, as discussed in the next chapter. This time, competitively speaking, I found myself on the same side as my former opponents. In my own defense, I think I used better tactics and found sounder grounds upon which to base a case.

Both the Big and Little Big Inch lines were privatized after the war and converted to carry natural gas by a new firm called Texas Eastern Transmission Company. More about that in the next chapter. Still later the Little Big Inch was reconverted to product service, and a new leg was built from Norris City to Chicago. By that time Phillips probably ceased to care. The old Great Lakes pipeline system was sold to the Williams Brothers, who converted it to a variety of uses.

Before we leave the Big Inch line, another problem should be mentioned. Before its construction, and with financing arranged, contractors engaged, and the first projects about to be started, some of the railroads refused to grant pipeline easements to cross their roadbeds. I descended upon the head of the American Association of Railroads, James Pelly, for help, but he refused me. Somewhat brutally he said he would rather run the risk of losing the war than risk losing rail traffic to pipelines after the war. When I failed to persuade him, I wrote the Cole Pipeline Act, named after its congressional sponsor, William Cole, which gave the federal government the right to condemn rights-of-way for the new federal pipelines. Congress passed the Act on July 1, 1941. Once having the power of condemnation, we didn't have to use it. We always negotiated our way through, and quickly.

Transporation issues took up much of our time with the press, the industry, and Congress.

Dealing with the Rubber Shortage

An almost overnight shortage of natural rubber for civilian tires, airplanes, and military vehicles hit us early in the war. All manner of "pie-in-the-sky" schemes were put forward. Missions were sent to the jungles of Brazil to find wild rubber, plans were advanced to extract latex from milkweed, an obscure desert plant, and massive programs were organized to collect used tires and bath mats. The Office of Rubber was created in September, 1942. A rubber "czar," William Jeffers, was assigned to the problem. As so often happens in Washington, the czar knew almost nothing about the problem. In civilian life he had served as the chief executive of the Union Pacific Railroad. Only the staff of the PAW possessed any expertise about how the enormous wartime demands for rubber might be met from synthetic sources.

To aid Jeffers, the refining division of the PAW first suggested what became known as the "quickie" rubber program. Aside from the fact that it was neither very quick nor of much use, it did point the Rubber Division toward the ultimate solution—once again by mustering and combining the technologies and cooperative efforts of the refining branch of the petroleum industry.

Butadiene was known as a potential source for the manufacture of synthetic rubber. The Germans had perfected a process, and Standard of New Jersey, as a participant in various I. G. Farben enterprises, possessed the know-how to do it in the United States. Butadiene was a by-product of certain oil refining operations and could also be synthesized from various oil and gas sources. The keys to accomplishing this were antitrust clearances to permit the pooling of technology, some governmental financing, and purchase contracts for a definite period of time and at some kind of a reasonable price. All of this was done in less than two years. It gave birth to the whole synthetic rubber industry, which has been with us worldwide ever since.

Poor old Standard of New Jersey took its usual public beating. Frank Howard, a vice-president who was Jersey's representative with I. G. Farben, was castigated in the media for having "withheld" vital technical information pursuant to pre-war agreements between Jersey and Farben. No one ever stopped to thank Howard and his company for having the know-how, which they were charged with hiding. If they hadn't been involved with the European Chemical Group, it is unlikely they would have had any information at all. As it was, they had it and made it freely available, which helped win the war by ending the rubber shortage.

The Challenge of 100-octane Gasoline

The rubber shortage was the forerunner of a far more serious potential shortage—100-octane aviation gasoline used to fly the great armadas of fighters and bombers needed on every front. The PAW saw the shortage coming. Our refiners had the technology but not much current capacity—only a few thousand barrels a day—in the face of a future requirement that would approach a million barrels a day. Pooling different patents, exchanging feedstock, and joint efforts by competing companies on a scale never before attempted was clearly indicated. After all, crude oils don't naturally contain 100-octane gasolines. Various molecules present in different crude oils in different proportions had to be torn apart, cracked, reformed, reassembled, and blended to make 100-octane material. Two essential steps were required—catalytic cracking and alkylation.

In the 1920s, refiners learned how to break up heavier hydrocarbons into lighter ones by thermal cracking, a process using high heat at high pressure. By 1940, catalytic cracking promised to do this same job more cheaply, at atmospheric pressures, and at relatively low temperatures, which resulted in a much superior yield of the components needed for high octane fuel. Eugene Houdry, a brilliant French scientist, had devised one method of catalytic cracking. He had a plant on stream at Sun's Marcos Hook facility. It worked but was so complicated that it took an army of skilled operators to keep it on-line. The process was not continuous. Duplicate units were built side by side. You ran one side for a short time, shut it down to burn the coke off the catalyst, and fired up the second unit to production. The catalyst was cooled with liquid metallic sodium—not exactly an environmentally approved material if it gets loose.

Another method of catalytic cracking—the fluid process—had not gotten much beyond the laboratory at this time. The process was continuous. It relied upon grinding the catalyst into a fine powder, mixing it with the material to be cracked, and then separating the catalyst from the cracked material after the reaction took place—this allowed us the use of the catalyst over and over again. In 1941, no commercial-size plant using this fluid process had ever been built or operated. A consortium of national and international oil companies had designed and patented the process. By this time, improvements to the Houdry Process had been designed to make it continuous. Instead of a fixed bed catalyst to which the hydrocarbons were exposed in alternating sequences, a chain lift of buckets, carrying solid catalyst beads, was designed to expose the beads to the material to be cracked. After the reaction took place, the chain lift carried the beads out of the reaction zone, the coke was burned off, and the buckets took the beads back into the reaction zone to do their job over again on a fresh charge of uncracked stock.

These two processes—fluid and fixed bed—were played off against one another. The PAW either "permitted" or "compelled" (depending on how you looked at it) the pooling of patents and know-how. No one was sure whether either or both processes would work in a

One of the large, complex refineries that supplied the petroleum products for victory.

commercial-size plant. It is always a big step from a pilot plant to a full-size facility. Yet the war would not wait; we had to gamble. We built both kinds of plants in about equal capacity. They both worked with almost no "bugs," and before the war was over, we were making a million barrels a day of 100-octane fuel. We needed every barrel of it.

Alkylation was also almost brand new in 1941. It made higher octane material than cat cracking. In nontechnical terms, it was the reverse of cracking. It took light hydrocarbon molecules, frequently gases, and put them together to make heavier liquid material with an extremely high octane rating—so high that when the military wanted 115-octane fuel, as they did late in the war, this required almost pure alkylate. There were different ways to make alkylate. Some had patented a process which used hydro-fluoric acid as a catalyst; others used sulphuric acid. The PAW forced the pooling of technology. Both methods worked, and both are used to this day to make acceptable high octane leaded and unleaded motor fuel.

One of the PAW's greatest challenges was the 100-octane program. The troubles were more economic than technical. Since it took government contracts for an adequate period of time at the right price to induce refiners to build new plants on a large scale, we had to bargain on behalf of the Defense Supplies Corporation and Defense Plant Corporation, both run by Jesse Jones, as to prices and terms. As fast as I got one group to agree to lower royalties regarding cat cracking, I simply made the other group meet it or lose the business. Loud were the cries and arguments as to why one process deserved a relatively higher royalty than another. Here I wore only my legal hat. Shamefully, perhaps, I denied all knowledge about technical matters, simply insisting they meet competition as to price and terms. Since each company was worried about its postwar competitive position, they all fell in line, one by one.

When it came to building cat crackers and alkylation units, naturally the majors argued they ought to build and own them all. Given the right kind of contracts with the government, their treasuries were big enough to finance the plants. For those few who were cash short, Jesse Jones stood ready to loan them money at a "high" 4 percent rate. True to his banking background, he insisted on 4 percent even though interest was a pass-through expense chargeable to the other branches of the same government in the price of aviation gasoline. As Jesse once put it to me, no one was ever going to charge him with lending the government's money at less than fair market rates. I suspect it is too bad that none of his successors ever seemed to share his views. They prefer to pretend that they are not giving our substance away when they make loans below the market rate of interest—such as is done by the U.S. Treasury to the federal power marketing administrations—or for nothing (or less than nothing after allowing for losses)—such as with defaults on loans made by the Export-Import bank.

Actually, when our major-company friends offered to build and finance 100-octane facilities on their own, such offers were not as generous as they appeared. Under the tax laws of the time, the companies could depreciate their total investment against income in five years. Most of the 100-octane contracts with the government ran for this period. They had a built-in profit which, theoretically, would cover all their costs. Thereafter, again theoretically, they would have a valuable producing asset on their books at no cost. Lest critics accuse us of overreaching, let me hasten to point out that we deliberately set up this system to get the plants built in a hurry. It is better to bait the hook with profit incentives than to try to subsidize and police a program of governmental building and ownership, with all the bureaucratic delays and inefficiencies bound to follow.

So much for the majors and the 100-octane building program. What about the independents? Here the Defense Plant Corporation paid for and owned 100-octane units built within the confines of or immediately adjacent to existing independent plants. The independents supplied, built, and operated these units. Most gambled that, by the end of the war, they would be the only logical buyers if the units still had useful life. In the PAW we thought so. We felt justified in not leaving all the high octane facilities exclusively in the hands of big companies when peace came (despite a PAW staff largely drafted from big companies). As expected, the majors were critical, and the independents built us no monuments. Back in civilian life as an independent, I once more met myself coming back on this issue. But that tale comes later.

We labored several months on the first model contract, one to build facilities at a refinery then owned by Cities Service near Chicago. The pricing clause for the purchase of 100-octane gasoline to be produced included an algebraic equation. I presented the contract to Jesse Jones, the head of Defense Supplies Corporation, the buyer of the gasoline. When he came to the algebraic equation, he looked up to ask what it meant. In jest, I told him it was derived from a favorite dirty limerick of mine. When I repeated it to him, Jesse laughed and signed the model contract. Neither the limerick nor the model contract bore repeating. The model was changed so many times, in fact, that later versions bore little resemblance to the parent.

What we did to stretch our supplies of 100-octane was almost unbelievable. To make just a little more, we had to shuttle blending stocks clear across the country by tank car. Cost (at ten dollars a gallon) was no object, but supply was questionable. When I heard about some of our troops in the desert filling drums with sand and 100-octane fuel to build a fire and keep warm, I shuddered. Stories of flying sea planes around at high altitudes in the tropics to cool a case of beer made me wonder how much that must have cost. Ten dollars a gallon for a lot of gallons was surely on the low side! But then, no one ever claimed that wars or any nation's military establishment were anything but wasteful. Our enemies no doubt could tell similar tales. Moreover, their high octane fuel had to be synthesized in part from brown coal (a low-grade lignite) rather than crude oil. The costs of going this route stagger the mind. After the war I walked through some of the German plants which made both gasoline and butter substitutes out of brown coal. Some of those in Congress and the energy bureaucracy in the 1970s and early 1980s—who spoke so blithely of making "synthetic oil" from shale or coal—should have taken a long, hard look at the costs of making it. As it turned out, they found out the hard way and at the taxpayer's expense.

To make the maximum amount of 100-octane gasoline, we had to take the goodies out of civilian motor fuel. We required additional blending of tetraethyl lead (an antiknock compound) and cut the octane number of regular and premium grades. For some, like the Sun Oil Company, this represented a competitive hardship. Their "Blue Sunoco" brand had always been advertised as lead-free, but now they had to add lead to make a marketable product. Chairman J. Howard Pew told Ickes and me that we were putting Sun out of business. The secretary "cried" with him but added that he still expected to see Sun doing business at the same old stand. They did; in fact, they liked lead so much—as the best and cheapest route to better octane numbers—they never took it out. Some said Sun expected their Marcus Hook refinery to blow up when the first tank car of lead arrived, but it survived.

Before leaving the 100-octane program, I want to record an interesting incident involving the Soviets. Midway through the war, the Soviets requested (and the president agreed to) a protocol whereby we consented, among other things, to provide them three catalytic cracking units. They asked for fixed-bed Houdry units. Bruce Brown of the PAW and I tried to tell them that such units were already obsolete. They persisted in refusing the far more efficient fluid and thermal catalytic cracking (TCC) units, and we couldn't figure it out. Bruce finally remembered that only the old Houdry units had ever been written up in the technical journals. They suspected we were trying to dump second-rate units on them—so violent were their suspicions—and so short were we of the equipment necessary to fabricate our new and better units—that we finally relaxed and let our allies have the old Houdry units. Bruce and I always wondered if they ever kept them on-line, since we barely could. The night that Sun and Houdry celebrated the Soviets' signing for a paid-up license, DeGolyer and I watched J. Howard Pew of Sun drink toasts to the Soviets with water. De and I did the honors with vodka. This gave us the courage to "liberate" a couple of gold plated coffee spoons from the Mayflower Hotel as a memento. I still treasure mine.

Flexing Our Muscles

After Pearl Harbor, we did away with the fiction that our requests of the petroleum industry were merely "recommendations." Such requests became directives, and they were followed "voluntarily" for the simple reason that the PAW controlled the supply of all critical materials and all operating supplies needed by almost every operator, large or small, in the business. Unlike the 1970s, when the DOE issued edicts and retroactive interpretations like bolts from the blue, each of our directives—and, later, each more formal order—was drafted and generally agreed upon by experts before issuance. Both the government and the industry knew ahead of time precisely what needed to be done and exactly how we were going to do it. The contrast between then and the 1970s—a period in which the industry and government were at odds—is described in a letter of March 22, 1979, which I wrote to the secretary of the Department of Energy, James Schlesinger.

Although directives without the force of statutory law behind them were enough to manage the industry, they were not enough to manage everything that needed managing—politicians, for example.

The Texas Railroad Commission, the proration authority in that state, was, of course, composed of politicians. Again and again we sought to persuade that commission to prorate and allocate Texas oil to provide the quality of crude—in the proper quantities at the places where it could best be used—we needed for the war effort. It did no good to allocate quotas for sour (high-sulphur) crude in far West Texas with no way to move it to the Gulf, where no refining capacity could handle this crude type. Politics being politics, the commissioners had constituents in far West Texas; thus, the politicians often felt compelled to do what they knew made little economic sense. To bring this and other similar situations in line, we needed something more than mere directives or "recommendations." We set about getting it—a delegation to the petroleum administrator (and his deputy) of all the powers of the president under the Second War Powers Act. These powers were so great that only the administrator's own discretion limited what he could order respecting any matter relating to petroleum at home and, through American companies, abroad.

After the usual Washington horse trading with other agencies, with each faction seeking to maintain its own jurisdiction unimpaired, an executive order was drafted and signed vesting in the petroleum administrator almost complete power over the petroleum industry and over all the other branches of government seeking to deal with that industry. It had never happened before—government now had authority over all oil and gas activities concentrated in a single governmental agency. The Department of Energy only pretends to concentrate such power roday, when in actuality the DOE itself is a house divided; authority over oil and gas is still spread all over downtown Washington and Capitol Hill.

On the evening of the signing of the executive order, we held an impromptu celebration in the DeGolyer-Marshall apartment. We hung a copy of the order on a venetian blind and offered toasts. In poker terms, we had drawn a pair of deuces and now held a full house: we were "legal" at last in our own right with our own statutory basis.

DeGolyer, who had always handled our relations with the Texas Railroad Commission as director of the Conservation Division, asked for a copy of the executive order—seals, ribbons, and all. He took off for Texas to tell the Railroad Commission how we needed to have oil allocated. Under the executive order, we could have taken over most of their powers, if we had had to. We had no desire to take over, but they feared we might. All we really wanted was for them to do what we needed to have done.

De described what happened before the commission when he returned. He asked the TRC if he could speak off the record and proceeded to tell an off-color story with an "I'm in charge" punch line. De then held up the executive order, and the chairman asked how we needed the oil to be allocated next month. After De submitted a schedule, the Commission chairman, E. O. Thompson, announced, "Dr. DeGolyer, it will be just that way." As with our condemnation leverage over railroad right-of-way, wise men were careful not to invoke our wrath.

In California, as always, there were no state laws according allocation to any administrative body. We dusted off our old procedures, which had been effective, but only for short periods under the NRA. We worked with the old Central Committee of California Oil Producers. We got them to remove most of the inequities in their schedules, which originated in their efforts to buy some more-or-less "voluntary" compliance and generally clean up their schedules pursuant to orders under the Second War Powers Act. Those few producers on the West Coast who wanted to pick their neighbor's pockets under the law of capture were made to toe the mark. Incidentally, it was the last time they were brought to heel; after the war, California reverted to its "law of the jungle" practices. By then, however, many California fields had been so abused that there was not much more damage that could be done to them.

One privilege that was purposefully not included in the delegation of powers to the PAW was the power to fix domestic prices for petroleum and its products. Such powers were left with the Office of Price Administration (OPA). I'd had my fingers too badly singed in the price-fixing fire once before to ever want to get directly involved again, particularly with a staff drawn from the oil industry. Nevertheless, I think we could have had the authority if we had asked. Some of our oil staff wanted it—no doubt for good, logical reasons. It would have made some of our work much easier, since price was (and is) such a powerful economic lever. Yet no agency with a strong group of oil company draftees could really use it without running the risk of being charged with trying to fix things up for their industry, their friends, or former company associates.

One of our PAW division directors, who came from a large organization not known for moderation in its political maneuvering, took up hours of our staff meetings demanding that we sponsor a petition to the OPA in favor of a general crude-oil price increase. He argued it would increase our capacity to produce more crude oil. Theoretically, he was right. Practically, since our bottlenecks were insufficient pipe and equipment to put into new wells, an increase in price would add little or nothing to our capacity. If we could get enough pipe, we could sustain enough capacity during the war with existing prices. Indeed, some thought we had so much excess capacity that we needed no more drilling.

I had to meet this argument once before a congressional committee. One member held what we called the "just go crack a valve theory," meaning that all we had to do was open valves in the oil field to flood ourselves. I told the worthy congressman a story of the East Texas oil field. Once upon a time the Texas Railroad Commission shut down the entire field's twenty thousand wells for a short period. Then they opened a few wells for one hour, multiplied their production by twenty-four and extrapolated that figure by the total number of wells in the field. This "proved" that the field had a capacity of some 130-odd million barrels a day. In theory this would have exhausted the remaining estimated reserves of the field in less than a couple of months, as compared with decades.

There was another power point in addition to oil pricing which the PAW deliberately avoided—civilian rationing of gasoline to stretch our scarce supplies. A few times we came close to getting involved with this practice, such as in the case of reduced wholesale gasoline deliveries and limited hours for service stations. Beyond these, we backed off. It was much too hot to handle. If there is any more politically sensitive nerve than the supply and price of gasoline for civilian use, we never found it. Fortunately, at least for us, the Office of Price Administration relished the job of rationing everything from tires and coffee to gasoline. Such rationing justified their jobs, salaries, local committees, enforcement divisions, lawyers, and, last but not least, their political patronage through favors to be granted to their own cadres of teacher's pets. As with later generations of oil and gas industry regulators (similar people in similar positions of potential power), they viewed their position as one from which they could almost run the country—or at least reshape it more in line with their own economic predilections.

One doctrinaire regulator of the OPA—particularly hostile to raising oil prices for any reason—was John Kenneth Galbraith. In mid-1941, the OPA convened a meeting with California producers, who were protesting an order to roll back their recent price increases. After the industry presented non-contested evidence to cost-justify the increases, Galbraith stunned the audience by stating that he had already decided the issue and the order would stand. Charlie Jones of the Richfield Oil Corporation of California could only yell, "Bullshit."

The OPA had no better success at rationing than anyone else. When anyone ever says that coupon rationing worked during the war, that person never watched it in operation. In almost no time flat, we had coupon inflation—more coupons than gasoline—and favoritism, corruption, black markets, and futile enforcement efforts. The regulations proved utterly unworkable with constant and complete breakdowns in whatever systems were instituted. Only one thing worked: letting the service stations in an area run dry. Then the PAW, the supply agency, would order limited amounts of gasoline to certain key distribution points to take care of hospitals, war workers, and police departments. The rest had to go hang until the ration boys could start all over again.

Area rationing began in May, 1942. Every car owner was entitled to an "A" card, good for three gallons per week or around forty-five driving miles. More essential users got a "B" cards good for eleven, fifteen, or nineteen gallons per week depending on need. An "X" card was carte blanche to the lucky holder—which, not surprisingly, included many politicians and bureaucrats. When area rationing turned into national rationing in late 1942, more categories were added. Naturally the task of classifying degrees of essentialness was a "baby" that even Solomon with all his wisdom could not have divided.

When the gasoline lines hit twice in the 1970s, coupon rationing was readied on a standby basis. A new breed of bureaucrats wanted to rush in to make the same old mistakes all over again. As has been said so many times, those who do not understand history are doomed to repeat it. More production at almost any cost, whatever the profit margins, is cheap compared to almost any system of governmental rationing.

I co-chaired the interagency Petroleum Requirements Committee on behalf of the PAW. We tried to balance the requests of the military, defense industry, and civilian claimants. Civilians got the left-overs, which the OPA rationed. After we told the OPA what that amount was, which initially determined the value of ration tickets, I always wanted to find a cyclone cellar until the storm passed. Once, I refused the military's request for a large allocation of heavy fuel in the state of Washington unless they told me why they wanted it. They took me to see Secretary of War Henry Stimson, and he briefed me on the Manhattan Project. They got the fuel. I wished I hadn't asked its purpose for fear I might talk in my sleep.

Throughout the life of the PAW, the old-line bureaucrats at Interior regularly sought to get their hands on our upstart administration. The PAW lived in Interior but was no real part of it. Once, when Ralph Davies, the deputy administrator, was home in San Francisco for a much-needed rest at Christmas time, Abe Fortas, the under-secretary of Interior, had his boys draft a secret order turning over the East Texas and Wilmington (California) oil fields to government operation. Under the delegation of powers from the president to the secretary of Interior, Ickes could have signed such an order. If signed, it would have completely disrupted our relations with the oil industry. A great friend of mine, then Interior Assistant Secretary Oscar Chapman, tipped me off after I promised him that I had to find the order myself and not disclose that I had learned about the proposed order from him. That was easy, since all government departments leak like sieves. I had a copy from an "unauthorized source" in less than an hour.

With Davies out of town, I had to beard the lion in his den alone. In private I reminded Ickes that he had promised to return the oil business to the same competitive status at the end of the war as he found it when the war started. If he condemned the two largest oil fields in the nation, the oil industry would regard his action as the biggest double cross in history. I told him I would have to agree with such an industry judgment, adding that most of us, including his deputy and chief counsel, in the PAW would have no choice but to walk out, and the administration's ability to lead would be destroyed. The secretary looked up, grunted, and told me he would think about it. Early the next morning, he telephoned. His statement was short and sweet—two sentences: "Howard, I think you are right. I am not going to sign that order."

Perhaps the whole incident was just another illustration of the reach for power always present in governmental agencies. As Ickes himself once remarked, "There is one position in the Department of the Interior which is never vacant—that of the assistant secretary who wants to be secretary." In this case it was an under-secretary, a personal friend of mine whom I had helped to elect to the position of editor-in-chief of the *Yale Law Journal*.

Afterwards I told Fortas in no uncertain terms to keep his cotton-picking hands off our oil business and, in return, we would keep ours off his Interior business. We had no more trouble from this particular quarter, but it seemed there were always some individuals or agencies trying to get their hands on the oil business. Looking at Washington during the energy crisis era, this only got worse with a multitude of agencies, bureaus, departments, and commissions asserting jurisdiction over oil and gas. The creation of the Department of Energy in 1977 was supposed to end all that, but I see that the DOE's *The National Energy Strategy: Powerful Ideas for America*, a 264-page document released to much fanfare in February, 1991, sought the advice of twenty-three other federal agencies!

Petroleum Administrator Ickes and Deputy Administrator Davies had coordinate powers—either could sign any order or regulation with equal legal effect. This meant that if either was out of town or unavailable, our business could go forward with the other in charge and post-facto without delay. No ratification or approvals of orders were required. This administrative set-up inadvertently served an amusing end. In wartime, only the head of each administrative department was permitted a car and driver. We were clearly an administrative agency, which meant only Administrator Ickes would have a car. Ickes had one as secretary of the Interior, and the car was known in the department as the "egg wagon," since it was sometimes used to deliver eggs sold from the secretary's farm. But, because in our case the administrator and the deputy administrator had coordinate powers, I was able to argue successfully that we had not one but two heads, and this bit of legal hair-splitting got Davies a car.

The Secretary was inordinately proud of what he called "my petroleum staff." He had a right to be. In later years I boasted that with this staff I could run most major oil companies better than they were being run.

It was nothing short of tragic during the 1970s energy crises that the politicians and old-line bureaucrats rendered us utterly unable to assemble the kind of staff we had in the PAW. But then it was not a

At a PAW dinner honoring Deputy Administrator Ralph Davies (*left to right:*) Marshall, J. R. Parten, Davies, and Justin Wolf.

wartime emergency, rather, it was a crisis of regulation, where none was needed.

Of course, not everything was sweetness and light when dealing with all members for the oil industry on all occasions. A long-time independent oilman on our staff was fond of leading us in a cheer by asking the question, "What makes the oil industry tick?" To which he and the rest of us would respond by rolling our eyes to the ceiling and chanting in unison, "gr-e-e-d and avarice." After the laughter died away, we would all confess that, for the most part, we were simply describing the nature of free competition. The hook has to be baited with the chance to make a buck if things are to get done.

Activity on the Pricing Front

It was only when greed and avarice went beyond normal bounds that we had to try to restrain it. I remember one such case which bothered me. A contract with the government to buy 100-octane gasoline in the Middle East came across my desk. When I saw that it involved

Standard of California and the Texas Company, I wrote across the file, "I deem myself disqualified," since, in the prewar days, Standard had been my client. Others in my legal division approved the contract and passed it through to Secretary Ickes for his signature. Later he called me to his office. He had the contract and file, and he had noticed I'd disqualified myself. He was nice enough to say that he had known me long enough to know that, in his eyes, I needed not to have disqualified myself. Would I please read the file and advise him personally whether he should approve it? I took it home to review, and afterwards I had to tell him not to. The price of the gasoline was pitched on the posted price of sweet, East Texas, forty degree gravity crude plus the full transportation rate from East Texas to the Middle East. In fact, the gasoline was to be made from sour Middle Eastern crude in a refinery located in the Middle East. As I put it to the secretary, "This is a basing point gone mad." Moreover, these Middle Eastern crudes were being sold to Japan just before the war for a mere pittance compared to the purely artificial crude oil prices included in this contract." The secretary's eyes gleamed. "Let's send the contract up to the Truman Committee," he suggested.

I knew a ruckus would follow. We had enough troubles without taking on a media battle that would create a diversion hindering rather than helping our war efforts. While I was disqualified to talk to my old client, Standard, to get them to rewrite the contract, I asked permission to go talk to the heads of the Texas Company. The secretary agreed—reluctantly, I think, because he always enjoyed a good ruckus.

A short journey to New York found me in the head offices of the Texas Company (now Texaco). I explained to the chairman and president how damaging it would be to their company, the oil industry, and the PAW if this bit of "overreaching" was aired by the Truman Committee. I took the position that I didn't want to write a contract for them; but I also asked if they would review this one personally to come up with a price formula that could be defended. They agreed, and I happily went on my way.

My rejoicing was premature, however, as a couple of days later one of the senior partners from my law firm in San Francisco—from which I was technically still on leave—called to report that the Texas people had complained to Standard, charging me with spoiling a great deal they had made to sell 100-octane gasoline. Thus, I was forced to violate a rule I had set for myself when I entered government service—never to intervene in matters affecting my former clients. I flew to San Francisco, told the whole story to the top executives of Standard, was thanked for doing what they thought was proper, and was told they would "handle" the Texas Company. They did, and the original contract was withdrawn. A new one was submitted with the price of crude oil, based upon actual sales of Middle Eastern crude just before the war, and adjusted slightly upward to compensate for increased costs since that time.

Except for negotiating prices to be charged for aviation gasoline sold to the government, the PAW steered clear of outright price-fixing for oil and its products. That task was deliberately left by us to the Office of Price Administration. This was a sound political decision, even though it was bad economics. After all, supply and price are opposite sides of the same coin—something that few of the well-intentioned idealists with the OPA ever understood. It was said in our shop that the OPA fixed prices that were "fair and equitable, whichever is lower," and that their economists could always "draw a bright line of logic between a set of false assumptions and a set of preconceived conclusions." Or as Garth Young, one of my later associates at Signal Oil & Gas, put it: "They could hurry quickly from the minor truths to the major fallacies."

Once the OPA came to the PAW and asked us to write an order compelling refineries to produce more heating oil and less regular gasoline from the same barrels of crude oil. I told them I could write such an order, but I also told them that neither we nor anyone else could either enforce it or really know whether it was being obeyed. The proportions in which a barrel of almost any crude in most refineries can be cut up are much too variable for any outside observer to ascertain. I added that, if they wanted the proportions changed, all they had to do was raise the price of distillate in relation to the price of gasoline, and it would happen automatically. Shocking as it seemed to the OPA to bait a hook with money, they did, and it worked. In the 1973–75 period, the Federal Energy Office, a predecessor agency to the Department of Energy, faced exactly the same problem and issued a "refinery-tilt" order instead of relying on economic incentive. The results were predictably bad, and new programs were relied upon until deregulation in early 1981. How little we learn from history.

I suspect our decision not to ask for direct pricing authority in the

PAW was wise. Drawn from the oil industry as so many of us were, we were constantly importuned to argue for higher crude oil prices. One prominent, successful, and wealthy independent, George Hill, regularly delivered highly emotional speeches on the subject of higher crude prices at meetings of the Petroleum Industry War Council.

People in the oil business gave these kinds of diatribes in public life as well. One in particular, Don Knowlton (drafted from Phillips Petroleum), served as director of production in the PAW. We rarely had a staff meeting without Knowlton urging us to sponsor a general crude oil price increase by the OPA. We never did because it was too touchy a subject and because we could not convince ourselves that it would do much good. At that time price was much less of a factor increasing oil deliverability than adequate supplies of pipe, manpower and other critical materials needed to drill wells. But Don kept arguing anyway. He was a petroleum engineer—a good one—but he thought that the oil business not only started at the wellhead but ended there.

One day a petition arrived from independent producers in the Borger area of West Texas. The producers claimed that the price of Borger oil had been frozen at relatively lower prices than similar oil nearby. Hence, the price of Borger oil could be raised without raising the prices of products refined from such oil. Of course, under this theory, those who bought and refined Borger oil would have to eat the increase.

Poor Don Knowlton apparently did not know his own oil company, Phillips, was a net buyer of Borger crude. They would have to do the eating. He assumed that any crude price increase was good, but this case was special. We let him recommend the increase to the OPA. Some of us ran around the flank to persuade the OPA to let this one go through, and they did. We sat by and waited, but we didn't have to wait long. Pages and pages of protesting wires were sent by Phillips to the OPA, the petroleum administrator, the deputy administrator and everyone else that their executives could think of. I was called by both the administrator and the deputy inquiring what all the fuss was about. I handed them a proposed reply, which read: "Suggest you take this matter up with D. R. Knowlton upon whose recommendation this price increase was granted." Later, I learned, Don had a bad week. I confess it was a rough way to teach Don (and, perhaps, his former company) an elementary lesson in petroleum economics. At any rate, it worked. For the rest of the war, Don dropped the subject of crude oil prices. Here was a perfect example of the advantages of having a staff in the government that understood the integrated character of the oil business and how to handle even their friends in that business.

Political Pressures

Even in the midst of World War II, politics never took a holiday. We were constantly harassed by the media and congressional committees. The media always wrapped themselves in the flag and screamed about "the people's right to know," regardless whether or not such knowledge would also be highly useful to our enemies. Congressional investigations, then as now, often related more to the reelection desires of individual senators and congressmen than to any real desire to learn the facts.

Drew Pearson, the famous Washington columnist, was my pet pigeon. When he kept calling for information which we knew he would distort, the deputy administrator would turn him over to me. Rarely have I ever used a recording device on my telephone and never have I done so without telling my caller in advance. But I always recorded my conversations with Drew. I wanted him to know that if he misquoted or misrepresented anything I told him, I would be able to prove him a liar, and after we came to that understanding I had no trouble. Unlike the professional politician, I learned the hard way that my job was easier if I kept out of the papers.

Then, as now, congressional hearings were much more a matter of playacting than serious investigations. Generally speaking, they found most of us in the PAW too hot to handle. Certainly Ickes was, Davies became so, and most of the rest of us learned how to play the game. When a senator tried to corner J. R. Parten, a rangy, confident, articulate Texas independent oilman with the PAW at the time, Parten stood tall and started to cross-examine his interrogator. "Senator," he said, "what do you know about the oil business?" "Nothing, Major Parten, nothing," replied the senator. "Well, well," laughed the major, "in that case, do let me explain it to you." And he proceeded to do so, much to the embarrassment of the senator.

At this same hearing another representative, obviously trying to establish a conflict of personal interest, went after Stuart Coleman, the director of worldwide supply for the PAW. He asked Coleman where he had come from. Stuart quietly mentioned Standard of New Jersey. What had been his position? "A vice-president." Now the killer question: "How many shares of stock in Jersey do you own?" Answer, in a low voice: "Fifteen shares." The room exploded. All the questioner could think to say was, "Well, Mr. Coleman, with that big an interest, isn't it either a buy or a sell?" Even with all the cards in their own hands, congressional committees sometimes came off second best.

Politics being politics, administrators are constantly required to make speeches, and the PAW was no exception.

Probably only Secretary Ickes enjoyed it. I once told him he was only tough on paper or behind a microphone. There he had no equal. Sometimes I had to pinch hit for him if he was unable to keep an engagement. Because I am allergic to reading prepared formal speeches, either my own or someone else's, I often started by holding up the secretary's manuscript, telling the audience it had been filed with the press and now I would try to tell them what the secretary would have said. Frequently, I had written much of the formal speech anyway, so it was not too difficult to give without recourse to notes.

Bruce Brown, an alumnus from Standard of Indiana who headed the refining branch of the PAW, together with some of the rest of us, evolved a test to measure when an American businessman had spent too long with the government in Washington. It was quite simple. You do not have to rise very high in the hierarchy to have to make speeches. After you have made one, you are entitled to think you have said something important. People come up to tell you so. But, if you think it is important the next morning, it's time to go home. If the sound of your own voice from a podium sounds impressive—beware! As for me—if I never have to write, deliver, or listen to another speech, it will be too soon.

Winding Down at the PAW

After 1943, the majority of the most difficult problems of the PAW were behind us. We had a smooth running government organization coordinating with expert industry committees, which blanketed the nation and the main branches of the petroleum industry—arbitrarily defined as production, transportation, refining and marketing. For marketing, "distribution" would have been a better name. You did not have to market a "shortage"—actually you had to engage in "unsalesmanship"—the art of convincing customers they really didn't need as much as they would like. President Roosevelt once jokingly complained to Ickes in my presence that we took fuel oil away from the overnight steamers plying between Norfolk and Washington. "In my younger days," he said somewhat nostalgically, "we called it the girl and bottle line."

Marketing was our simplest problem. We just cut off materials for building or expanding distribution facilities almost everywhere. Peacetime free competition had built excess capacity in service stations, bulk plants, terminals, tank cars, tank trucks, barges, and small product pipelines. We proceeded to use up our excess capacity, which, though wasteful in peacetime, was useful in wartime. We forced exchanges to eliminate cross and back hauls; we diverted barges, trucks, tank cars, and pipelines from short hauls to long hauls in the right directions. We integrated the industry as it might have been under an ideally run monopoly. "Brand" names were ignored, and the advertising of more or less imaginary superiorities of one company's products versus another went by the boards. Sun put tetraethyl lead into its lead-free "Blue Sunoco" gasoline, and the Texas Company had to take "Fire Chief" gasoline from refineries other than its own. It is probably fair to say that the consumer never knew the difference. During the gasoline lines of 1974 and 1979, the consumer was much more concerned with getting the stuff than about its color, brand, or the service that went with it.

My own experience under the Oil Code in the 1930s kept me from falling into the trap of trying to regulate "marketing practices," though we were constantly panhandled to promulgate rules which might help one marketer or class of marketers at the expense of others. As might be expected, these pressures came even from within our own marketing division, where some of those who had spent their lives in marketing had trouble readjusting their thinking.

In refining, particularly with the 100-octane program, the PAW and the industry distinguished themselves. Starting almost from scratch, by the close of the war we were producing around a million barrels a day, which converts to 42,000,000 gallons a day or 1,260,000,000 gallons a month. In addition, many of these gallons exceeded 100-octane by at least fifteen numbers, the quality demanded by the best aviation gasoline engines as the war progressed.

On the oil-production front, after the initial struggles with states like Texas, California and Illinois—and after the pirates who, under the law of capture, tried to drain (steal) from their neighbors were brought to heel—matters went smoothly enough. By and large, crude was produced in the right amounts, at the right qualities, and in the right places to once again float us to victory on the proverbial sea of oil. What it really cost us as a nation, in terms of our future supplies of low-cost oil, we were not to learn until several decades later. For, while we seldom exceeded (at least by wide margins) the maximum efficient rates of production in our "flush" fields, there can be no question that we were throwing our low-cost reserves around the world, both during and after the war, as though there were no tomorrow. This is something our politicians and people at home do not seem to understand. It is easier and better politics to damn the domestic oil industry—while our erstwhile Allies conveniently forget the role the same companies played.

Now the war was winding down, and I was physically and financially exhausted. Sixteen hours a day, seven days a week—the strain of constantly recurring crises coupled with the then-minute governmental salaries wore most of us down. At least it cured me of smoking. When I hit five packs a day, I decided there must be a smarter way to kill myself. Somehow I mustered the nerve to quit. Had I not, I doubt if I would be here to tell this history.

In the early fall of 1944, Secretary Ickes and Ralph Davies were persuaded to let me go. I had served the OPC and the PAW as Chief Counsel from July 25, 1941, until October 20, 1943, and as Assistant Deputy Administrator from October 20, 1943, until my departure on September 1, 1944. It seemed a long time since 1941, when I had agreed to come for six months to help get the old Office of the Petroleum Coordinator for National Defense started.

I had a replacement for the chief counsel's office. They no longer needed a man who wore two hats as chief counsel and assistant deputy. Strangely enough, my legal replacement was none other than Robert Hardwicke, one of the leading oil and gas lawyers in Texas, who helped straighten out two young whippersnappers from Yale when first they came to the oil patch to double-check a monograph on oil and gas they intended to publish. He was a far better and more orthodox lawyer than I could ever hope to be.

Some of the legal eagles in the big law firms probably breathed a few sighs of relief with my departure, and I can't say that I blame them. In my own self-serving defense, I can only plead that if they had always followed my (and their) orthodox legal training, I don't know how we would have ever gotten the wartime task done in the time allotted us. Cutting red tape and legal corners never really bothered us. Throughout this time, we were never sued or enjoined, and we were never in court, except once or twice when we appeared as a friend of the court. We had a minor enforcement division best known for its "inactivity," primarily because we got compliance, since we knew what we were doing, and the industry knew it. In the rare instances when force was called for, our control over the critical materials, which everyone had to have to stay in business, was quite enough to bring any recalcitrant into line.

Often during the war, many of us were told by leaders in the oil business that they would be forever grateful for the job we were doing and for the manner in which the PAW operated. Such compliments were usually accompanied with words to the effect that, when we returned to private life, if there ever was anything they could do for us, we had but to ask. Privately I did not believe such statements when they were made, nor did I believe any honest corporate executive could or should grant any of us any special future favors to compensate us for doing the kind of job they and we believed in. Regarding my own "treatment" after the war, I was right. No special consideration was ever offered by any of those who spoke of it. Only one special advantage—an important one—has ever accrued from my years of government service: if I had a problem or a deal to present, I could usually get a hearing on the merits of it.

As with most departures, last lunches and dinners are customary, and there sticks in my memory a last lunch in the private dining room of Secretary Ickes, attended only by those three who immodestly regarded themselves as the creators of the PAW and the managers of the whole affair. After a few martinis and a little brandy, toasts were in order. After Ickes and Davies had properly toasted and roasted me, it was my turn. Mine was short: "I appreciate the compliments you both have paid me and your regrets at my leaving. I want to say that for three years I have run interference and cut down tacklers for two of the greatest running backs I have ever known. As for me, I now have nothing but bruises all over my body and am glad to go home. Here's to both of you."

CHAPTER FIVE

Building Ashland Oil & Refining, 1944–51

After Washington it was back to San Francisco, perhaps to practice law. As any lawyer knows, when you leave a law practice for three years, upon your return you have to build a new one, since, while you've been away, others had to take your clients. My own situation proved even more complicated.

The chairman of Standard of California, Harry Collier (who'd suspended the retirement rule for himself), regarded Ralph Davies as an enemy and potential competitor for his position. Collier had good reason to fear Davies. Before the war, old Harry was guilty of persuading the powers that be to reverse the board's election of Davies as president in order to seize the position for himself. Because of my association with Davies, Collier informed my partners that he preferred I not represent Standard until such time as he might decide otherwise. My partners asked me to accept this verdict because the firm had plenty of other law to practice. They also thought it would not be long before Collier would want me again to represent Standard, given my own oil and gas experience.

Leaving Pillsbury, Madison & Sutro

All this gave me reason to ponder. I remember the skillful approach my friend Pete Jones of Cities Service took toward some of our key personnel in the PAW. He suggested that, after the war, they should seek employment through him. When he approached me in the same manner, I suggested he cease and desist, as he might be seen as soliciting special consideration from PAW personnel.

Paul Blazer of Ashland took a different approach. A few months before I negotiated my release from the PAW, he invited me to lunch. He told me he was searching for a successor when it came time for him to retire as president of Ashland Oil and Refining—he wanted to make himself chairman and bring in a younger man as president. Could a particular individual we both knew handle that job? I had no trouble with a negative answer. The individual in question was a competent producer but his knowledge of the oil business ended at the wellhead. Ashland was much more of a refiner-marketer than an oil producer. Did I have any suggestions? I asked what he was prepared to pay, and he named a figure. In jest rather than with any serious thought of seeking an offer, I told him that, at that price, I ought to apply for the position myself. Too quickly he came back with, "would you"? Later, when I got to know Paul pretty well, I was sure he had set the trap for me. The other chap about whom he inquired was a way of opening the subject he really wanted to talk about. A couple of times later in my own business, I borrowed Paul's approach, and I found it invariably worked, although I never tried it on a government employee.

At the time Paul tried it out on me, I cut off the conversation. He was told that as long as I was with the government, I would not discuss leaving it with him or anyone else in the oil industry. Now, however, I was a free man again. I picked up the telephone and called Paul Blazer in Ashland, Kentucky. Of course he recalled the luncheon. Did he still need a president? He did and asked how soon I could be in Ashland. That question was answered a couple of days later after a short session with the head of Standard. Collier, now sixty-eight, directly confirmed what my senior partners had already relayed. To acquiesce to Collier's decision meant severing my association with the oil industry until he might change his mind. Obviously I could not represent competing oil companies as long as my firm represented Standard. Standard's interests were so broad that dealing with almost any other party would represent a conflict of interest. I decided on the spot. I think I smiled, and I hope I kept my voice down, when I said: "Mr. Collier, you cannot cut me off from the oil industry, or get rid of me as you did Ralph Davies. You aren't big enough. I think my own stature in this business is as good as your own. It's nice to have known you. Our paths may cross again." It would have been better to leave without those last words, but I failed to follow old Oscar Sutro's admonition: "Never get mad except in cold blood."

My swan song probably resealed the fate of Ralph Davies, who returned to San Francisco when he left the PAW in May, 1946. Not even a highly publicized visit by Harold Ickes to Collier to plead Ralph's case did any good. Ralph knew the end had come and formally resigned from California Standard on the eve of Ickes's visit.

Leaving the directors' floor of the old Standard Oil building, I took the private elevator (sometimes known as the subpoena escape hatch) one floor up to my law firm. I called Paul Blazer, reached an agreement on the details of becoming president of Ashland, and regretfully resigned my partnership in Pillsbury, Madison and Sutro. As someone later remarked—heretofore I had practiced law with my right hand and the oil business with my left. From here on it would be the oil business first and the legal profession second, which was, in my case, a distinction without much practical difference except for courtroom appearances.

My New Company

Within a week our bags were packed, and our San Francisco house was offered for sale. My wife and two young boys were bundled into an ancient Buick to set out cross-country for Ashland, Kentucky, two thousand miles away. I still remember the magnificence of the fall coloring as we drove the last forty miles out of the hills to the valley of the Ohio River. We all wondered what would befall us in a brand new life with a relatively strong, but young, independent refining company.

We were born and raised in the East, had learned the far West, but were coming into the Middle West. Kipling was only partly right in proclaiming that East and West would never meet. In oil, competition is much the same wherever you find it—only the actors change. So it was with Ashland Oil and Refining. Economically, there are no hard and fast lines between east, west, or middle west in the oil industry. They are all inseparably interrelated. What happens in any one general area promptly affects all the others, and only under monopolistic conditions could it be otherwise. As long as competition is relatively free—and in oil it is more than just *relatively* free—oil and its products flow quickly to equalize any shift anywhere in supply and demand.

ASHLAND OIL & REFINING

Ashland Oil & Refining Company stands as a rare example of a subsidiary which swallowed its parent. When the shallow oil fields of eastern Kentucky were first found and developed in the 1920s, an independent producing group, the Swiss Oil Company, headed by the enterprising J. Fred Miles, found itself with a surplus of crude oil. This company discovered, as many have, that the mere finding of oil does not automatically result in Cadillacs and Rolls Royces driving up to the door the next day. Oil is worth only what someone will pay for it—if they choose to buy it. The Swiss Company, certainly not of Swiss origin, might just as well have looked to Switzerland for markets as for buyers in the backwoods of eastern Kentucky. Not only were there no decent roads in this area, there were no accurate road maps or land surveys either. Only a few old timers with Ashland had any idea who owned what and where that what was located. To me, accustomed to land maps properly surveyed in sections, Ashland's lease maps in eastern Kentucky looked like a jigsaw puzzle.

Confronted with a marketing puzzle in a frontier backwater, the Swiss Oil Company decided, as many of their predecessors had before them, to get into the refining business. They needed cash for their theoretically valuable but presently worthless samples of crude oil. They asked a young man, who was only seven years removed from an advertising agency in Chicago and who lacked any formal training in the technical side of petroleum distillation, to get them into the refining business. His name was Paul Blazer, and he'd left the advertising business to sell petroleum products and had picked up refinery experience along the way.

Swiss Oil was exceedingly lucky, for Paul was a superb salesman, and what he initially lacked in technical background, he learned the hard way. Indeed, he learned it so well that at the time of his death he was probably the best and most versatile refiner-marketer in the whole oil business—major or independent. When it came to making a refinery profitable, he was without peer. Max Fisher, starting with fifteen hundred dollars, built the Aurora Gasoline Company. He too learned refining on his own. He was fond of complaining about refinery engineers as technically competent people who had to learn that the purpose of a refinery was to start crude oil at one end, take products out the other, and make dollars in-between. Paul Blazer never had to learn that lesson. His eye was always on the dollar.

Blazer acquired an unreliable one-thousand barrel skimming plant on the banks of the Big Sandy River near Ashland, Kentucky, on January 26, 1924. The Catlettsburg, Kentucky, plant had been a money loser for a series of owners. It did business under the name of Ashland Refining Company, a wholly owned subsidiary of the Swiss Oil Company. Swiss's crude surplus was moved to Ashland by a moderately-sized (at least for those days) pipeline, which connected with other lines extending to the East Coast. Within several months, after some improvements and a staff reorganization, it became a money-maker. The next year, Blazer's Ashland Refining Company did much better than its parent. So bitter were the disagreements that board meetings of the parent sometimes ended in fist fights. They finally agreed to disagree and merged Swiss Oil into Ashland Refining to become Ashland Oil & Refining Company, effective October 26, 1924. They turned the whole operation over to Paul Blazer, and he proceeded to run it, as he ran everything else in his long career, successfully. He always bowed, but only as a formality, to the concept of corporate democracy and the power of a board of directors, though he actually dictated—and should have, in my opinion. In the final analysis, Blazer's rule always came down to "all those in favor say aye, all those opposed submit their resignations." No nonintegrated, independent oil company on the way up in any branch of this business can build on a committee system, whether that system operates within the company or on its board of directors.

Ashland always made money—even in the depths of the depression when so many others, including some of the majors, either showed red figures or went to the wall. When I came aboard in the fall of 1944, we were running some twenty thousand barrels a day in that simple refinery. Strangely enough, some of the original equipment was still in use. Old batch stills, hooked in series for continuous rather than batch operation, sometimes were run to clean up odds and ends of off-specification material. These old stills looked like museum pieces—but they passed the acid test by making money.

Ashland had long since outgrown its simple beginnings when I arrived. The original skimming plant had been modified and expanded to include a sizable crude unit, thermal cracking and vacuum facilities, an asphalt plant, and a host of auxiliary facilities and tankage. Under wartime contracts with the Defense Plant and Defense Supplies corporations, Ashland built a thermo-for-catalytic cracking plant on nearby land to make 100-octane aviation gasoline for the military.

The headquarters of Marshall's new company as he found it in 1944. *Courtesy Ashland Oil, Inc.*

This unit had a rated capacity of seventy-five hundred barrels a day of light distillates, which could then be converted into high octane gasoline components. With related facilities, including an alkylation plant and its auxiliaries, Ashland had a wholly modern refinery complex able to compete with anyone in the business.

Ashland operated but did not own the complex, since the Defense Plant Corporation paid for it and held legal title. How to acquire that title was one of my first major assignments.

Despite crude runs to stills of well over twenty thousand barrels a day in 1944, we only had a few thousand barrels of "captive" crude supply, which we either owned or controlled in eastern Kentucky. This limited supply flowed down an ancient pipeline system Blazer had acquired. All the connections that once allowed oil movements east of Kentucky had been abandoned, which left Ashland as the only ready buyer of oil from hundreds of small but long-lived "stripper" wells in all the eastern Kentucky fields. Small trucks traversing dirt roads, and tank cars moving on secondary branch lines, cannot compete against a largely depreciated pipeline.

About four thousand barrels a day of this Kentucky crude flowed down our old pipe. The economics worked for Ashland because our out-of-pocket costs for pumping the oil were small, the oil itself had some good lubricating components (higher priced products), and we did not have to pay as much for the oil, relatively, as we did for Illinois Basin crudes hundreds of miles to the west. As might be supposed, our pricing policies in the fields of eastern Kentucky hardly qualified us to win a popularity contest with these oil producers! In the old days, the "wicked" Standard Oil Company, which dominated the pipeline sector, was always accused of such tactics; in this case an independent was on the receiving end of such accusations. But if we paid more for eastern Kentucky oil, there would have been an insufficient refining profit. If we had bought elsewhere, our Kentucky producers would have had no market at all. Producers never want to grant this, and who can blame them? Competition has nothing to do with abstract "justice" or the "inherent worth" of "my" oil. A refiner tries to buy as cheaply as he can; a producer always believes the "fair" price of his crude is more than he is currently receiving. Unless government sticks its fingers in the pie, the market decides how the pie is cut up.

Ashland had far outgrown its local beginning. A few thousand barrels a day of eastern Kentucky crude amounted to only a few drops in the bucket of what was necessary to run a refinery that profited from economy of scale. Fortunately, another substantial oil province, the Illinois Basin, reborn just before the outbreak of the war, was accessible via the Ohio River where the Big Sandy emptied a mile or so below the refinery. Shallow Pennsylvania oil sands had yielded impressive amounts of crude years before, and many wells were still "stripping." Not much more was expected, but, as so often happens, new exploration tools (in this case the reflection seismograph), coupled with "wildcat" drilling by speculators, led to a succession of prolific new oil fields from deeper horizons. These horizons were still relatively shallow by modern standards, and, as in California, no conservation laws, rules, or regulations stood in the way of any operator drilling on any spacing pattern he pleased or producing any well as fast as he pleased—regardless of the effect on ultimate recoveries. Once again, this was the old law of capture at its worst. Centralia, Louden, Uniontown, Pure Ridge (so-named because it was first found by the Pure Oil Company), and many others were rushed into production in a wild frenzy of town-lot drilling and wide-open production. Once more the night skies were lit with burning flares from the gas associated with oil. You had no time to waste if you were to produce both your own and your neighbors' oil before they got theirs and yours too.

Bad as this might be for most oil producers, as short-sighted as it might be in wasting the oil and gas resources of the nation, it was made to Ashland's order. The Illinois Basin covered southern Illinois, southwestern Indiana, and western Kentucky, and most of it fronted the Ohio River. Ashland, Sohio, and others moved quickly to lay pipelines to terminals on the river or connect lines to refineries located in or near the Basin. Crude oil—light and relatively sweet—still went begging. Ashland plunged into the barge business, and it built its own barges and tow boats (though these were actually prime movers that pushed rather than pulled). Eventually, Ashland's fleet became the largest on the inland waterways. These tows were operated on "milk runs" twenty-four hours a day in both clear weather and in fog, after the invention of radar. A big river terminal at the mouth of the Big Sandy near the refinery pumped the oil off each tow in a matter of hours to send the tow immediately back down river for another load.

This river system was better, and much bigger, than our own pipeline in eastern Kentucky. The economics, however, were the same.

A big surplus of the high-quality crude—and our own physical facilities to gather and transport cheap oil at nominal rates to our own low-cost refinery—assured us a market for everything we could produce. In refining, it is axiomatic that you make the most money on the last barrels—unless you have to cut your prices too much on all your barrels to move those last ones. Because of our low costs from the wellhead through the refinery, we achieved a substantial competitive edge over most of our competitors in our natural markets. We always ran full-out, with excellent overall margins of profit. One of Paul Blazer's proudest private boasts was that he never curtailed any refinery because he lacked either crude or market. In public he sometimes advocated curtailing such runs, but he meant his competitor's runs, not Ashland's!

This sketch of the history and position of Ashland details how I found the company when I came down from the hills in the fall of 1944. A complete painting is provided in Otto Scott's *The Exception*. There was much more that I didn't know. For example, my new chairman had, at least tentatively and probably definitely, promised the presidency of the company to one of the vice-presidents before I came into the picture. It took several years for me to learn of this. It made life no easier in some areas within the company. I sometimes wonder whether I would have accepted the presidency had I known of the alleged commitment. Apparently the other candidate was shocked to his knees when I arrived. Now, after part of a lifetime in corporate affairs, I suspect this was not the first or the last time something akin to double-talk ever occurred in the rarified atmosphere of executive management. In government, we all came to expect it. In private life, on those occasions when I have been exposed to such double-talk, it has always come as something of a surprise and is always tinged with some disappointment.

Getting Started at Ashland

After a few weeks in a hotel room, I got settled in Ashland. I found a house, not among the socially elite on Bath Avenue, where my chairman lived, but out among the "Indians" (the streets were named after Indian tribes—our street was called Seminole), where younger executives lived. Every Midwestern town has a "Bath Avenue." We saw enough of each other, almost day and night, for seven years, even without being next-door neighbors.

ASHLAND OIL & REFINING

My first day as president was September 1, 1944, and I found myself quickly immersed in the details of Ashland's many-sided activities. They ranged from the company's limited exploration of the Illinois Basin, production both in that basin and eastern Kentucky, crude oil purchasing, river transportation of crude oil and refined products along the Ohio River, pipelines and pipeline movements, the intricacies of modern refining and, finally, wholesale and retail marketing. I found myself quite at home in exploration and production—really my first love in the business. Marketing—wholesale, and retail—was easy enough; I had been through almost too much of it. Refining, Ashland's most important money-maker, was something else. While I understood refining in general (meaning, I suppose, that I knew what was going on both chemically and physically inside a refinery), I had absolutely no experience running a profitable refinery. This I had to learn quickly, as my board chairman knew these things, and, as president of Ashland, I was expected to know them, too. I had the best teacher in the industry, Blazer himself, and learn I did. Since then, I've helped run a series of refineries which have turned a profit. The marketplace is always the final test, and the balance sheet marks the examination papers. Both, as they should be, are cold and completely impersonal.

Almost as soon as I was settled, we were hit by a nation-wide refinery strike called by the Oil Workers Union. This was to be only the first of many union strikes, work stoppages, and collective bargaining sessions through which I was destined to live over the years ahead as a refinery executive. This one was easy, for the war was still on. Some of us persuaded the Department of the Navy to "take over" all the refineries to assure ample supplies of "navy special" fuel oil. For some months our refinery was nominally run by a young, junior-grade lieutenant under orders from an admiral in Washington. For the sake and safety of the Navy, he did not really run the refinery—if he had, it would surely have blown up. But the fiction averted a strike which was settled, nationwide, on terms which I have forgotten.

A Detour for Reparations

In the spring of 1945, one last serious interruption in my transition to a private oil career from lawyer-oilman (and governmental lawyer-administrator) occurred. A telephone call came from an old client and associate, Ed Pauley of California, head of Pauley Petroleum. He was a Democratic politician par excellence and a personal friend of Harry Truman who helped engineer Truman's vice-presidential nomination at the Chicago convention. Pauley was an old legal client of mine, and a great one at that—always in trouble and highly solvent. I was indebted to him as well for his front work with FDR to establish the Office of Petroleum Coordinator.

Ed's telephone call was direct and to the point. He explained that Truman had appointed him ambassador in charge of reparations, and Ed wanted me to be a part of the delegation that was being sent to Russia. My answer was direct and to the point—"Pauley, I have just dedicated three years of my life to my country at $8,600 a year, and I'm broke. I'm trying to rebuild my shattered fortunes. I'm not going with you to Russia or anywhere else." His only answer was, "Well, we'll see."

I relayed the conversation to Paul Blazer and asked him to agree that, no matter what Pauley did, he was not to consent to my joining Ed as his lawyer on a mission to Moscow. Paul agreed, as he thought the idea was ridiculous anyway.

That night the phone rang in Paul's Ashland house. It was the chief operator of the Ashland phone company. In a breathless voice she told Paul, "The President of the United States wants to talk to you." Although I was not privy to the conversation, Paul reported his conversation as follows: "Mr. Blazer, we want to borrow your young man for a few months. Mr. Pauley wants him to serve as counsel for the American Delegation on Reparations to meet shortly in Moscow. I want him to take the appointment. Will you grant him a leave of absence without pay?" I am sure the "without pay" posed no problem, except for me. How did Paul reply to the President's request? "Yes, Mr. President, anything that Ashland can do." Paul asked me, "What else could I do?" I could only reply, "Nothing I guess."

A week later I was on my way to Moscow via the Azores, London, Paris, and Berlin with the Air Transport Command and the rest of the delegation. Back with the government, and this time with the State Department, I was paid a per diem of twenty-five dollars. The rule was that, if you could save any, you could keep it. In less than a week, I was shot with every known inoculation against most known diseases. The only disease I caught, however, was from a yellow fever inoculation. Yellow fever, I learned later, had a vaccine that had not yet been perfected. That knocked me flat—high fever, delirium, and all—for forty-eight hours. But the episode had a happy ending.

I mentioned in the first chapter that, as a child, a severe case of typhoid fever left me with a short left leg and almost no normal left hip joint. Thanks to both my mother and my athletic career, I made do with what was left of a badly damaged left leg. Just before I left Washington for Ashland, that leg began to hurt. Johns Hopkins Hospital prescribed a bone graft. Since I could not spare the needed months of immobility, I postponed what they advised was inevitable. I just worried along with things as they were. To my surprise, some weeks after the bout with the yellow fever inoculation, I noticed the hip no longer hurt. Returning from Russia, I checked with Hopkins. They confessed to a misdiagnosis. They explained, somewhat apologetically, that they now believed my trouble was bone arthritis in the damaged joint, for which the cure was an artificial fever. I had gotten the fever accidentally and was cured.

I joined my old client, Ed Pauley, at his office in the White House a few days before we left for Europe and Russia. As I walked in, I could not help but notice his wide grin at having "drafted" me, to which I could only say, "You old son-of-a-bitch." I was supplied a top-secret version of the negotiations at Yalta, upon which a definitive reparations agreement was expected to be based. The negotiations made me shudder. Churchill was the only negotiator on our side who seemed to have had any realistic appreciation of the problems. Stalin knew what he wanted—everything he could get—and my old idol, FDR, seemed to think he could charm the fangs out of tough old Joe. We were assigned the impossible task of trying to make sense out of an ambiguous and lopsided "agreement" on reparations. The whole story has been recorded by others, so there seems no reason to repeat it here.

Although this section represents a digression from the central focus of this book, a few sidelights are worth recounting. The reparations delegation was divided right down the middle between Truman appointees and Roosevelt appointees largely selected by Henry Morgenthau, then Roosevelt's Treasury secretary. This was exactly what the Russians wanted—in their case for purely selfish reasons. They figured to lay claim to everything that could be moved out of Germany. If the people were left to starve without the means of earning a living, so much the better. If we were concerned about the German people, they coldly suggested, "you support them."

None of us wanted reparations for ourselves. After the experience with reparations following World War I, Harry Truman's contingent definitely had no desire to force the United States to pay reparations indirectly to Russia by filling the vacuum left in Germany after the Russians removed everything they could lift. We watched Russian troops in the eastern zone of Berlin move everything that was not tied down and much of what was. This included the working parts of heavy industry, light industrial equipment, and all manner of household materials. If walls or concrete were in the way, the walls were knocked down and the concrete torn up. Half the rails from double-track rights-of-way were removed. In our office quarters in Berlin, the doors, windows, and toilet seats were already gone. We always wondered what they did with toilet seats. One of our fellow delegates remarked, "these fellows make Genghis Khan look like an amateur."

So deep was the division between the Morgenthau boys and the Pauley group, there was no way to have a strategy session without the Russians learning all about it even before we could meet with them. Among us, we had not one but many informers, perhaps well-intentioned but (seemingly) emotionally unbalanced. A good example of this was our second in command (who held the rank of minister) who would have headed the delegation if Truman had not appointed Pauley.

In the final analysis, Pauley won out, for he had a direct pipeline to the president. The opposition within our delegation seemingly worked for the Russians, who had no such division. Our group "won" but in a negative sense of the word. We learned how to stop the Russians from stripping West Germany as they had the Eastern sector. They exercised unilateral control in outright violation of the Yalta Agreement providing for a unified Germany. Ed Pauley and I agreed that these Russians made some of our "wicked" oilmen look like saints. Fortunately for the United States, Truman had the sense to select an ambassador who had fought his way up in a competitive business. No one, not even the Russians, could push Pauley around. Our diplomatic service could use many like him today.

In Washington during the war, I sometimes wondered where I would end up after the conflict, and a more unlikely place than Germany I couldn't have imagined. When Averell Harriman announced that Japan had surrendered and it was all over, I found myself attending a reception in our embassy in Moscow—standing between commander-in-chief of the Allied forces, Dwight D. Eisenhower, and commander-in-chief of the Russian forces, General Georgi K. Zukoff.

Some of us "bigwigs" on the reparations mission were flown from Moscow to Potsdam to sit in on some of the Big Three meetings. At the midpoint of these sessions, Churchill had to return to Britain for his reelection campaign which, alas, he lost. The temporary adjournment of the meetings called for a press release. Each delegation submitted a draft explaining what had been accomplished thus far. Since that was almost nothing, it had to be embellished with high-sounding phrases. Our draft was long, ponderous, inconsistent and loaded with dangling participles and split infinitives. Truman read it. Churchill looked across the table to say with a chuckle, "Mr. President, that may be in good American, but it's awfully poor English." I had to hide a smile. Churchill knew the difference, but the ghostwriters in our State Department, both then and now, are hardly masters of proper English.

After the meetings reassembled, Potsdam droned on. Pauley and a few of us on his staff were assigned to quarters outside of Potsdam alongside the house where Truman resided. One afternoon, Truman's military aide presented himself to Pauley with the words, "Mr. Ambassador, the president's compliments sir, and he says you are all two drinks behind." That problem was quickly cured.

Incidentally, all of us had "assimilated" military ranks, since all Germany was under military government. Pauley held the rank of lieutenant general. Along with a couple of others, I was merely a lowly brigadier. Our identification cards were stamped, "no good unless captured by the enemy." Perhaps so, but along with our "rank" of general officer, these cards gave us food, lodging, and transportation the like of which I had not experienced since before the war. A room at the Ritz in Paris and Claridges in London was two dollars a day. Meals, including steaks at the general officers mess, cost about the same. Johnny Walker "Black Label" scotch, a favorite of mine long since among the war effort "missing" at home, was a dollar a bottle. In the next war, rather than fighting the battles in Washington, some of us thought we would prefer to be a general officer almost anywhere. With such fringe benefits, you could almost live on that twenty-five dollars per diem—at least until we got to Russia, where there were no amenities, high fixed rates of exchange, and still higher prices for us outsiders. Only the mountains of fresh caviar and vodka served at Russian banquets and state dinners made life livable.

Buying Crude in a Seller's Market

With the end of the war, almost every economist in the oil industry confidently predicted an overwhelming surplus of crude oil and refined products, particularly gasoline. Relieved of the large demands of the armed forces, few foresaw anything but a fast return to the prewar excesses of supply, which had characterized the decade of the 1930s world-wide. One of those few was my chairman, Paul Blazer. Just why he felt there would be a shortage rather than a surplus, I'm not sure even he knew. Certainly no computer told him. No econometric model would have given him a clue, since it would have been programmed with inadequate data. Market sense, the seat of his pants, the intuition of experience—call it what you will—Blazer did foresee a shortage. Hindsight showed why he was right. Civilian demand had been restricted during the war. With the war over, civilian requirements shot up, not to where it might be expected to be after the war but to where such requirements would have risen had there been no conflict.

What a lesson for the arrogant forecasters among the academics and bureaucrats of today! They presume to tell us about future demand. If they overlook even one obscure factor, or make a single false assumption, their predictions are almost worthless. Put another way: "Experts hurry from the minor truths to the major fallacies."

Faced with a prospective shortage of oil and controlling only a little of our own, Ashland set out to buy crude for its existing and expanding refining capacities. Few today remember that crude oil was in short supply from 1945 to 1947. Too many looked upon the 1970s shortages of oil as something brand new for peacetime. The task of finding and buying crude in a seller's market fell on my shoulders. Mostly it was my job to buy, since Ashland never had an exploration department worthy of the name; even if we had had such a department, there was no way to find and develop the production to fill up our refineries. Besides, Blazer believed we could always buy crude oil cheaper than we could find it, and, for Ashland at least, this was probably true. Given its great expertise in refining, and its almost total lack of such expertise in production, all of Blazer's trained intuition surely whispered that we should concentrate our efforts and limited capital resources in fields where we were among the best as opposed to wasting shot and shell in an area where we were among the worst.

Many years later, long after I had left Ashland and after various unsuccessful attempts to become profitable explorers and producers of oil, another generation of Ashland executives apparently learned Blazer's lesson all over again. They sold almost all their oil production to raise cash for areas expected to offer better returns. This approach to the oil business ran directly contrary to that followed by major integrated companies and favored by most security analysts. Both rated integrated companies according to how much their refineries were supplied by internally-produced crude oil. Only their own reserves in the ground mattered. As one oil banker viewed it, "refining is just a bad habit."

Relying on the outside market proved beneficial to Ashland, particularly in one case. In 1978, the company took advantage of rising oil prices to sell their exploration and production subsidiary for $1.2 billion on the belief that the government would allocate crude to them if they were caught short. When President Carter banned Iranian exports to the United States shortly thereafter, Ashland threatened to close one of its refineries rather than pay forty dollars per barrel on the spot market. The Department of Energy stepped in and forced a group of companies to supply Ashland; the group included one company that had purchased Ashland's crude at a higher price than that at which it was now obligated to sell to Ashland. Oil price and allocation decontrol in early 1981 did not come a moment too soon.

For Ashland and me in 1945, no such comfortable governmental intervention was available to relieve us of the necessity of competing for crude oil—we had to buy or die. I discovered it took the same ingenuity and imagination to buy crude oil in times of shortage as it took to sell refined products in times of surplus. In each case respectively, if you paid too much or sold too cheaply, you could go broke. Our crude contracts were generally pitched on "posted" prices. We did not dare pay a premium over these prices for some oil without jeopardizing the prices we were required to pay for the bulk of our purchases bought under contract (or "open division orders"). Such orders could be terminated by any producer on thirty days notice.

Caught between needing more crude and not being able to risk a general increase in our posted prices, which would have been immediately met by our important competitors, how could we persuade independent producers either to change their connections or sell newly-discovered oil to Ashland? Fundamentally, it came down to a series of individual instances. What could you do for a particular producer not directly (or at least obviously) related to the price per barrel paid for oil? I found I had to deal with each individual producer personally as president of Ashland, unhampered by any committee system or board "approvals." It took me down the highways between Ashland, the Illinois Basin, and the Southwest to the tune of driving tens of thousands of miles a year. Later, a single "puddle jumper" eliminated some of the driving. But before this, there were numerous times when I would leave Ashland around six in the afternoon to drive four hundred miles to the Illinois Basin on curving two lane roads. I'd work all the next day with producers in the Basin, only to set sail back to Ashland in the early evening of that same day. In this competitive game, there were only "the quick or the dead."

Some of the deals illustrate how even today there is so much more to oil prices than appears on the surface (or can ever be gleaned by merely looking at "published" or "posted" prices). At this time, oil country tubular goods—casing and tubing to run in new wells—were scarce. In return for a "call" (or option) to buy oil at posted prices for a period of time, I agreed to furnish many a producer with needed tubular goods at mill prices. How much over mill prices I had to pay was my secret. Usually I didn't know—at least not exactly. When you barter fuel oil (then also in short supply) for pipe with steel mills, then move the pipe on a backhaul out of Pittsburgh on the decks of your own oil barges, how can even a CPA really figure your costs?

One prominent independent oil producer needed to borrow half a million dollars. He had negotiated with Sohio for a loan secured by his producing properties with interest at five percent. The deal was making its way through Sohio's formal procedures. For a "call" on all of his oil for five years, I offered him the same loan—with two "sweeteners." We would make the loan the very day after the mortgages, which on the record, were and give him a "side letter," which was off the record, waiving the interest on the loan. I still remember the producer's look of amazement. "You mean you will agree to this right here and now?" he asked. He wanted to talk to his wife. He did, and she agreed. "Call in your secretary," I said. "I am still a lawyer. I can dictate a letter agreement which we can both sign." We had a sound loan plus a

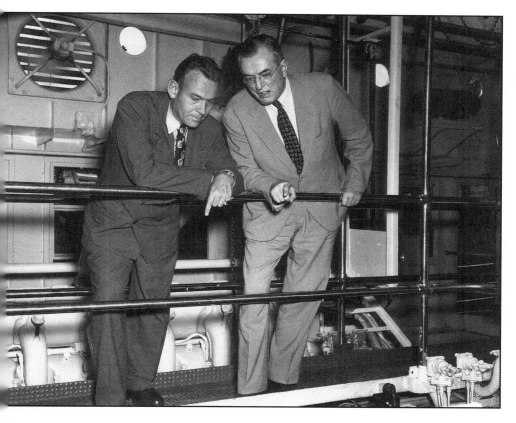

Marshall and Blazer inspecting the new river towboat, M/V Ashland, 1946. *Courtesy Ashland Oil, Inc.*

thousand barrels a day of crude oil, on which we expected to make a dollar a barrel. Later the top purchasing executive of Sohio asked me how I had out-bid them for that producer. I told him we just did it quicker, but it was a long time before I confessed to waiving the interest on the loan. Over the years, I loaned many millions of Ashland's money to independent producers without waiving the interest. For Ashland, these were relatively safe loans. The borrower pledged his oil to secure the loan and repaid the loan at so much a barrel for oil run through our pipelines and gathering facilities. We knew the producing properties and, unlike conventional financial institutions, needed no independent appraisals.

I missed once. In this case, the producer falsified his production records. After taking unmitigated hell from my chairman for a bad loan, good fortune let me turn the tables. When the loan was not paid, we foreclosed on the leases, from which repayment had been expected. Months later, some wildcatter stumbled into a good limestone producer offsetting these leases, and we quickly drilled four wells on the foreclosed leases. Instead of losing $200,000, Ashland ended up making a bundle—poetic justice!

I could not resist suggesting to my chairman that I knew about the prospective production from the start. The loan and foreclosures were simply the method employed to acquire leases that we could not otherwise have gotten. When he inquired ten years later whether I really knew about the production potential of these leases, I wondered at how he had swallowed my tall tale hook, line, and sinker. If Paul Blazer had known as much about exploration as he knew about refining, he would have known there was no way for me or anyone else to predict the productive whereabouts of the particularly erratic and tricky limestone in the Illinois Basin. One of my friends once drilled a well into this formation which produced over a million barrels. He followed it with four dry holes—boxing the compass north, west, south, and east on twenty-acre spacing right alongside the one big producer. Sometimes bad leases come back from the dead; sometimes good discoveries turn sour.

Building pipelines or gathering systems too early in a newly-developing area was another way of doing something "extra" for producers. "Too early" meant connecting the wells before you could see a payout from normal tariffs or throughput rates based on the then "known" reserves. Producers prefer pipelines to trucks—lines generally provide cheaper transportation to market and hence a better wellhead price. Pipelines are also less likely to be interrupted by weather, strikes, etc. We took many risks extending pipelines before they were economically justified. If payouts were sometimes slow, we at least bought much needed oil to make a refining profit.

It used to be said that an oilman's word is better than his bond. While often true, some of us in my generation insisted you had better know about the oilman who gives his word, and an example of this rule comes to mind from the Illinois Basin. In the course of chasing some oil from a new area, I had a handshake agreement with a very prominent independent oilman to the effect that, if we took the risk of laying twenty miles of pipeline to bring to market oil from some new wells his company had just completed, we would have a call on this oil for a period long enough to pay out the line at somewhat low (but agreed upon) rates. We bought his oil for about a year, when one of our competitors offered him a premium of fifty cents a barrel over "posted" prices. He accepted the offer, conveniently forgot about his word, and had the gall to ask me to allow him the use of our pipeline at nominal rates. We told him he and his new purchasers better hire themselves some trucks. This would have eaten up the fifty cents premium and then some. We kept the oil, but I made a mental note to get an ironclad contract in writing with this producer the next time we did business with him.

"Deliberate" losses from acquiring overpriced or poor prospects from independent producers were another device for buying oil. After a reasonable length of time, the leases could be abandoned ("quit claimed") and written off, depending on our tax bracket. This allowed Uncle Sam to help pay our cost of buying oil—which may or may not have been offset by taxes which the favored seller presumably had to pay. Here again we sometimes outsmarted ourselves when, occasionally, a bad lease turned out to be good. Neither the seller of poor leases, nor we the buyer, could always be sure which leases would not produce, any more than we could be certain which leases were certain to produce. This added spice to the pudding.

So many and so varied were the schemes and devices invented to compete for the right to purchase oil that no one has ever set them down. They ranged from estate planning, property settlements during divorce proceedings, entertainment at the Kentucky Derby, "free" legal services, discounts on products refined from part of a producer's oil—in fact almost any legitimate favor or assistance we could provide at a justified cost to ourselves. They were limited only by the ingenuity of sales-minded people, lawyers, accountants, and tax experts.

In all my years with Ashland, we never ceased having to compete for crude oil. With additional and expanding refineries, our requirements increased constantly, and before long we moved into areas outside eastern Kentucky and the Illinois Basin. Again we laid what our competitors thought were uneconomic pipelines—once as far as a hundred miles down the center of the Sprayberry trend, which was then a big (but questionable) new oil field in West Texas. Pipe was so scarce at the time that we had to dig up an old line in western

Pennsylvania, which had been in the ground for decades. Holes in the pipe were spot welded, old screw joints were cut off to convert the pipe to modern welded connections, and the reconditioned old line was trucked all the way to Midland, Texas. The payout was protracted, but much-needed good-quality crude was purchased.

How was oil brought home to Ashland's refineries in Kentucky and Ohio from Texas, Louisiana, and the Gulf Coast? Some moved through the truck lines of our competitors directly to our river terminals in the Middle West. Still more was exchanged with other companies owning refineries on the Gulf or in Oklahoma. They replaced (or exchanged) the oil we delivered to them with oil which they controlled closer to our refineries, with the price adjusted for quality.

Exchanges are sometimes portrayed as monopolistic, though nothing could be further from the truth. They allowed us to compete for crude over wide areas where it otherwise would have been impossible. Likewise, even though pipeline ownership has frequently been condemned as restraining competition, we almost never failed to get crude oil home either directly or through exchanges from these lines. Only rarely does a competitor owning a pipeline have you by the throat. In eastern Kentucky, where we owned the only pipeline outlet, producers thought we had them in a corner—and honesty compels the admission that, for the most part, we did. But we spent a lot of capital (which producers didn't spend) to build the line (and much more) to refine their crude. If one of our competitors, with whom we exchanged or over whose lines we shipped, brought enough oil in eastern Kentucky to make it worth their while, we would have felt ourselves compelled to exchange or transport down our lines. As has been said before, the whole market for oil and its products moves together almost everywhere, and the foregoing is just another example of this.

To gain the opportunity to buy just twelve hundred barrels a day of eastern Kentucky crude, Ashland bought out an obsolete sixteen hundred barrel a day refinery owned by the Texas Company (deep in the hills of eastern Kentucky) at a place called Pryse. With it came its supply of eastern Kentucky oil, which we could take down our lines to Ashland. Actually we ran the refinery temporarily on sour crude moved by tank car from west Texas. Normally, such a movement would have made absolutely no sense. But the government was short of "Navy special" fuel oil and paid a price that allowed us to move and refine this sour crude. We let the sulphur content of the crude damage some of the refinery equipment at Pryse, since we knew we could run it only so long as Navy special was in short supply—a matter of months.

One amusing incident marked this refinery operation. During the war, the government rented five 80,000 barrel tanks in the refinery yard to store two hundred proof, chemically-pure ethyl alcohol to be used in manufacturing synthetic rubber. The government removed all the alcohol, except about an inch's worth, in the bottom of each 80,000 barrel tank. The refinery boys solved that problem with sponges, rags, and jugs. Diluted with water, the alcohol was highly potable, and there was more than enough to keep our crew drunk for months. I tried some myself, and it tasted about like good vodka—far better than the Kentucky moonshine advertised as "at least one month old, colored and flavored with wood chips."

By the lucky accident of being able to run the refinery to make "Navy special," we almost got our purchase price back. When we shut it down, we moved a sizable amount of usable equipment and transferred about five thousand barrels a day of royalty-free thermal cracking capacity to Ashland. Despite protests from Universal Oil Products, our thermal cracking licensor, this saved us a nickel a barrel on thousands of barrels for quite a few years. Strangely enough, the lawyers for the Texas Company never realized they had sold us a free license with the refinery. We didn't realize it either until a few days after the purchase, when I just happened to be reviewing the contract on another question.

After calculating in the "extras" arising out of the refinery purchase, we earned the right to buy that twelve hundred barrels a day of Kentucky crude for no premium at all. I often wondered if the Texas Company ever understood why we bought their old plant or what they had sold. As one of my independent friends commented: "Some of us little fellows can live pretty well on the crumbs that fall off the tables of big major companies. I hope the trustbusters never put them out of business."

None of us in Ashland invented "the oil payment." That invention came from the fertile mind of Al Meadows, the long time head of General American Oil Company. Al used oil payments as a means of purchasing oil producing properties, but we adapted it as a financing device to assist us in buying oil at the wellhead from independent producers. Since an oil lease is regarded as "real" property, a producer could carve out a piece of the real estate in the form of a fixed number of barrels. This carved-out "oil payment" was sold to a shell corporation (a corporation without assets) which pledged the oil payment to a financial institution as security for a loan. The shell corporation then paid the producer for the carved-out oil payment with repayment tied to the pledged barrels. In this way the loan was made without recourse to the producer's personal credit.

For arranging such financing for producers, Ashland took a call on the right to buy all or part of a producer's daily production for an agreed-upon period of time. If the producing properties covered by an oil payment were carefully appraised, and could produce enough oil to liquidate the oil payment plus interest, a lending institution was not actually at "risk." Ashland bought millions of barrels of oil by this route. There was only one catch. The oil payment could not be "guaranteed" by Ashland, for as the tax laws then stood, such a guarantee converted the oil payment into a loan rather than a "sale." Only a sale could be cost-depleted to avoid taxes on the dollars flowing back.

When I started on the trail of crude purchasing for Ashland, I was a hopeless amateur. But since necessity is always the mother of invention, after some years I became a professional. I ceased to fear that we would be unable to purchase sufficient crude oil to run a profitable refinery. Before, as a major oil company representative, I feared marketing a surplus; now, as an independent, I learned how to buy to avoid a shortage. It is an interesting business.

As part of my attention to the small wells that fed our refineries, I represented Ashland as president of the National Stripper Well Association in 1946. Perhaps more than any integrated industry executive, I knew the thin economics of the small producer.

Securing a Cat Cracker

Turning from competition for crude oil to competition in refining, it was obvious to the better independents that the days of the small skimming plants were numbered. Gone were the days when a small refiner could survive by merely boiling some crude oil and separating whatever came forth into gasoline, kerosene, heating oils, and boiler fuel. With the advent of thermal cracking (and later catalytic cracking), efficient refineries became petrochemical plants. They tore apart the heavy molecules in crude oil to make higher proportions of more valuable light products, like gasoline. Then they recombined the molecules of light gaseous hydrocarbons to make high octane liquids, like alkalate or polyermized gasoline. They learned how to blend butane, normally a gas, back into motor fuel to make more and better quality gasoline.

To remain competitive in an era of advancing technology, Ashland needed a catalytic cracking unit with all its related facilities. Such a unit in full operation stood right alongside and tied into our principal refinery. Alas, we did not originally own our unit. We arranged to build and operate it for the Defense Plant Corporation to make 100-octane gasoline for the military. The federal subsidy amounted to $16,562,940, the third largest in the 29-unit, $236 million program. Here again, I met myself coming back from the other side of the industry.

In my PAW days, coincidentally, I had played a major role in providing a means whereby independent refiners could participate in the 100-octane production. Otherwise, almost all the best refineries would have ended up exclusively in the hands of major companies. Without governmental assistance, they alone were building, operating, and, more importantly, financing sizable cat crackers.

Once the war was over, the Defense Plant Corporation offered their crackers for sale. The sixty-four dollar question, as usual, was price. The majors argued for a price based upon cost less only normal depreciation, which usually resulted in a price so high that few independents could afford to pay it. Now I had to take on the task of arguing the independent's case before a senate investigating committee. Sometimes a strong attack is the best defense, and here that tactic worked. It was easy to charge our major friends with desiring to limit the ability of independents to compete against the majors, who owned advanced, efficient refineries. When the majors claimed that the independents wanted to buy too cheaply, I countered with the claim they had gotten their cat crackers for nothing.

The argument was simple: most of the majors' profits made on the sale of petroleum products to the military during the war were subjected to "renegotiation." Whatever they had to refund—and such payments were large—cost the companies one-hundred cents on the dollar. To whatever extent accelerated depreciation had been allowed on the plants built by the majors with their own funds, they saved more than a percentage of their income taxes. They increased their "costs" to diminish, dollar-for-dollar, refund obligations under "renegotiation."

When the hearings were over, I received a backhanded compliment from the president of a major company who had once tried to hire me and now protested that I knew too much—another example of why you can't expect competitors to love you. In this case my testimony did not destroy a personal friendship I enjoyed with my opponent both before and after the Senate hearing.

After the trial, Defense Plant Corporation finally decided to put its facilities up for bid. Since such facilities were really useful only to the adjacent independent refiner, the bidding was less than spirited. Ashland bid a little over two million dollars for the cat cracking facilities needed to compete against the majors. I don't recall any other bidders, except perhaps junk dealers. We didn't worry about such bids. Unless a refinery can be run at a profit, it is sold by the pound. It costs almost as much to tear one down and reerect it at a new location as it does to start with brand new equipment.

By acquiring a modern cat cracker with all its auxiliary facilities, Ashland had taken a giant step forward. Blazer still remembered the day when this plant was almost completed, as the construction unions tried to stretch out the completion. As Defense Plant inspectors walked through the plant, they noticed little work—instead, a series of crap games were in progress. Blazer told the inspectors that, if they did not like the stakes in a particular part of the plant, there were several other games in other areas where the stakes might be more to their liking. He added that, if any of the loafers were fired, the unions would see that they were back a day or so later. Crap games would not last, but union problems would remain.

A thermal catalytic cracker (or TCC) was the basic unit Ashland had bought. Future experience would show it was not as good as a fluid unit, but it was good enough for the late forties. Ashland modified its TCC from "bucket" to "air" lift. This helped, but when the time came to replace the unit, a fluid catalytic cracker (or FCC) unit was built. The same evolution followed with the Great Northern Oil Company, a refinery in Minnesota that I would cofound in the next decade, as discussed in chapter six. A TCC unit could be kept on stream for about a year versus a three-year turnaround for a FCC cracker.

An important note about refinery capacities can be found in Ashland's experience with its cat cracker. When purchased, it had a rated capacity of some seventy-five hundred barrels a day. When I departed Ashland early in 1952, we were running almost thirty thousand barrels a day. We had spent some millions removing bottlenecks, improving this or that, adding here and there, but nothing like the cost of a new thirty thousand barrel unit. From this comes a simple rule: refineries can usually be made to run substantially more than their "rated" capacities, depending on how they're operated, the product slate, and the kind of crude oils, or other material, fed to the plant. One able independent described how his refinery made money: "We run the plant so hard it almost blows up."

Purchasing New Refineries

Our old refining facilities were quickly tied into the new. The thermal cracker, vacuum unit, asphalt, and road oil plants—these older facilities could do things which the newer equipment could not. But tied together and integrated, a still better overall refining operation resulted. As the refined-product markets shifted either in terms of volume or price, we had the flexibility to change the relative production of the plants to maximize yields. From now on, refinery expansion came by the merger route.

Mergers, if carefully selected and negotiated, were a far less expensive means of expanding refining capacity than building new facilities from the grass roots. The lure of the acquired refinery was often the capital for modernizing it. In the case of privately-held companies, an additional attraction was public stock that was much more suitable for estate planning.

Mergers with Ashland followed in such regular intervals over five years, that I told our printer never to destroy the type on our most recent prospectus. The first of these mergers was with privately-held Allied Oil Company for a reasonably good refinery at Canton, Ohio, and a sophisticated fuel-oil sales division. The head of Allied was W. W. Vandeveer, a wartime friend from my last term of servitude in Washington. Van and his partner, Floyd Newman, were not getting along well. Both wanted out of the relationship—an ideal situation for a prospective (and favorable) merger. Ashland's problem was to merge without using money. A combination of voting preferred provided the sellers with a reasonably certain annual income, and the common had a good chance of future appreciation—this was the kicker in the deal. Since both stocks had voting rights, the sellers did not have to pay taxes on the appreciated value of their company until they sold some of their Ashland stock. The Ashland-Allied merger was finalized on August 3, 1948. Virtually overnight, Ashland almost doubled in size with Allied's pipelines, tankers, tank cars, crude production, and oil leases.

The formula evolved in the Allied merger became the pattern for a series of similar mergers, and, in succession, Frontier Refining in Buffalo, Aetna Oil in Louisville, and Valvoline at Pittsburgh yielded to this same general formula. I confess to a particular satisfaction in the case of Frontier. It too was run by a wartime friend in Washington, Jim Breuil. Standard of California, my old alma mater, had almost bought the company when Jim told me he wanted to sell. We used our Allied formula, added a few new wrinkles, and let Jim keep his producing subsidiary, Kingwood Oil Company. We ended up with all of Frontier's eight thousand barrel a day refinery and marketing as well as a subsidiary, Central Pipe Line, which gave us direct access to additional crude oil supplies in the Illinois Basin and the southwest. I was tempted to call old man Collier, still head of Standard of California, just to tell him our paths had crossed again. Now, I'm glad something saved me from being that stupid. It was more fun to laugh with Jim Breuil anyway.

Ashland securities were initially traded over the counter, then on the American Stock exchange, and finally on the "Big Board." Even Paul Blazer, who founded the company, never owned more than 10 percent of the common, and often his share was less. But with innumerable friends as stockholders, all of whom knew he'd made them money, he effectively controlled Ashland as if he owned it lock, stock, and barrel. One cardinal rule he always observed in any merger: "No deal is any good," he said, "which jeopardizes our operating control of this company." If such a risk offered a voting trust, Paul Blazer as trustee solved that problem. In effect, such trusts—the biggest of which Paul executed at the end of the Allied negotiations—lasted until a big-enough block of stock was distributed to the public to remove any threat of operating control by outside interests.

All of our refinery additions were acquired to advance Ashland's scale economies. Aetna, at Louisville, and Valvoline, at Pittsburgh, were both on the Ohio River. They could be served by our own barge fleet for inbound crude oil, outbound refined products, and the transfer of unfinished, semirefined products from one refinery to another. Our anchor refinery in Ashland could upgrade material from our other plants and supply high-octane blending stocks to improve the quality of gasoline originating in our other plants. Although the Allied plant at Canton did not exactly fit the Ohio River pattern, it was a large supplier of fuel oils to the steel industry in Ohio and western Pennsylvania, and the refinery at Ashland often needed these outlets. Light petroleum products can always be moved at a price, but many independent refiners have found themselves drowned in their bottom-heavy fuels, which are sometimes almost unsalable at any price.

Ashland often got rid of some of its bottoms at better prices than it received for gasoline, which was usually regarded as the "money crop" of refining. We produced a variety of high-quality asphalts for roads and roofing and tight-specification coal-spray oils and asphaltic binders for coal brickets. The last had to be very carefully made to avoid melting in a bin in the summer, hold coal fires together securely, and never smell up the house before being consumed for domestic heating. All of these products attested to Paul Blazer's ingenuity in refining. He did what our competitors either could not or failed to do. He even produced acceptable road asphalt from the residues of thermal cracking—an accomplishment which our major competitors had thought impossible; Shell even sent a technical team from overseas to find out how he did it, but I'm not sure Paul ever completely disclosed his trade secret.

Ashland's refinery acquisitions were all fundamentally based on Blazer's well-founded conviction that he and his crew could usually convert refineries owned by others into much more profitable operations. In private, Paul liked to boast that he could run any refinery better with half the labor force others thought they needed. Whatever he acquired on the basis of what they had been earning, Ashland ended up with probably the lowest capital cost per barrel of refining throughput of any company in the business. One refinery, which we bought for money rather than stock, was the old National Refinery of Findlay, Ohio. Paul merely walked through the ten thousand barrel a day plant for a few hours before we got it. His expert eye told him exactly what had to be done to transform a loser into a winner: "No wonder they have no residual fuel problem. They waste it running the plant. Some heat exchangers, better fraction items, and more asphalt production, and this thing can show a decent return."

The National purchase was unique, as it was the only refinery we ever bought for less than nothing. I was guilty of making the deal. The owners were teetering on the edge of bankruptcy, and I offered to pay cash for their inventories based upon a formula geared to assumed market prices—but only if they would throw the refinery in for nothing. I think they thought we would scrap the plant, but instead Paul merely shut it down for a half-million dollar renovation. It made money from the day he started it back up. When I finally sold off the inventory, the formula used to value such inventory was sufficiently conservative to leave us with half a million dollars more than we had allowed the seller. In a final, minor argument on inventory values, the sellers claimed fifteen thousand dollars more on a certain product than we thought was owed. It was a doubtful point, but I agreed to waive it. My chairman accused me of giving away Ashland's assets, and all I could say was that I didn't have the heart to take the last pound of flesh. Perhaps Paul Blazer had a weakness after all. He often skinned up the other fellow so badly that a wise man never came back for a second treatment or another deal. Personally, I wondered whether it always paid to go for the jugular.

Dodging the Big Inch and Little Big Inch

My experience in Washington, D.C., would come in handy during my Ashland years. My congressional testimony concerning Ashland's federally-subsidized cat cracker, as mentioned above, helped to ensure its future within the company. Around the same time (although this was incidental to Ashland) and at the request of Ralph Davies, I authored an executive order from President Truman to new Secretary of Interior Julius Krug, establishing the National Petroleum Council. With the demise of the PAW on May 3, 1946, the NPC came into being as the peacetime industry organization to advise the federal government in petroleum matters.

A third governmental matter of the period was very important to my company. At the end of the war, the Defense Plant Corporation found itself as the owner of the Big Inch and Little Big Inch pipelines that extended from the Gulf Coast to the North Atlantic. These lines were no longer essential to move crude oil and refined products to the East, as was the case during the war, but they probably offered a cheaper method of movement, depending on their capital cost, than conventional shipments by tanker or otherwise. It's almost a rule of thumb that big diameter pipelines moving large volumes over long distances provide the lowest cost per barrel.

Since the federal government wanted no part of the peacetime pipeline business, the two lines were offered for sale to the highest bidder. Ashland had an important economic interest in the uses to which these lines might be put. Those of us who refined and marketed in the Middle West (just west of the Appalachian Mountains) were on the fringes of a market which remained insulated by the transportation differentials between the competition of big refineries on the Gulf and big refineries on the East Coast. These refineries were supplied by water-bound crude originating in the Gulf. We could look out our windows at Ashland to watch barges moving downstream with refined products, which had come from the eastern plants, passing other barges moving upstream with similar products refined on the Gulf. If the Big Inch or Little Inch lines were put into crude oil or product service, our transportation advantages might be radically altered. Standard of Ohio entertained similar fears.

An outfit calling itself "Big Inch Oil" organized itself to bid for the lines. Some who wanted no part of direct gulf-coast competition argued for leaving the lines idle for use in another emergency. At the time it seemed difficult to beat something with nothing. I was able to persuade my chairman and, strangely enough, one of our toughest competitors, Standard of Ohio, to put up a pot to campaign for the conversion of the two lines to natural gas transmission—putting to use a product flared and wasted in many of the producing oil fields throughout the Southwest. Our group became known as "Big Inch Gas."

A bitter political battle followed. It was waged, as such battles usually are, in the media and before congressional committees. I even persuaded my old boss, former Interior Secretary Harold Ickes, now in private life, to publicly support converting the lines to natural gas transmission as a conservation measure. I took the same position for the same reason in my testimony before congressional committees. We prevailed, I think, on the merits of our cause. That it helped protect the competitive position of Ashland and Sohio (in this case strange bedfellows) was never mentioned.

The winning bid for the lines, announced in February, 1947, came from Texas Eastern Transmission Company, a corporation organized by Brown & Root (a big construction firm from Texas), my old friend

Everette DeGolyer, and Holley Poe, the director of our natural gas division in the PAW. They figured out that so long as they did not price themselves out of the market for cheap gas on the East Coast, the more they paid for their lines, the more profit they would make. This apparently strange result arose from an incentive under public-utility regulation. In the sales agreement, the Federal Power Commission agreed to recognize the price paid to the federal government—in this case $143 million—as a capital cost. This became the rate base upon which a "reasonable" return to the common stockholders of Texas Eastern was based. The key was to borrow most of the cost of the line for an interest cost less than the return allowed on the total cost of the line. All of the difference between the total allowed return and the lesser financing cost—multiplied by the rate base—accrued to the benefit of the common. This was how Texas Eastern, one of the nation's large interstate natural gas pipelines, now a subsidiary of Panhandle Eastern Corporation, got its start.

For my help in bringing about the conversion of the Big Inch line, I was accorded the right to subscribe to a small block of the promoters stock in Texas Eastern at a dollar a share. I exercised the option, offering half the stock to Sohio, since they had paid for half the cost of our campaign. They turned it down. I think they feared ever having to admit they were involved in our campaign. Their economic interest in having the lines used for gas rather than for oil or refined products was the same as ours—but they were regarded as a major and Ashland was just a struggling independent. Of course I assigned all my stock to Ashland at my nominal cost. Later, Ashland sold its Texas Eastern stock for a lot more than we and Sohio had spent on our campaign. Had there been any honest way to hang on to my promotion shares, I would have. All I could do was to buy a few shares personally at the same (much higher) price paid by the general public. When I sold them years later, I made a few dollars but not a killing.

Labor Union Troubles

Shortly after joining Ashland, all "collective bargaining" was dumped in my lap. It never was a pleasant assignment, and it got increasingly difficult as we acquired other refineries, whether "organized" or "unorganized." The job seemed to take a month out of my life every year. Bargaining with the Oil Workers Union often reminded me of trying lawsuits. Their merits were incidental, as it was largely a political and dramatic exercise.

Union officials, both at the local and national levels, must regularly take new post offers home to their constituents. Otherwise, they may fail reelection. Moreover, it is naive to suppose that labor peace is really desired by those who run unions. What really serves their interest is trouble, regardless of whether it is real, imagined, or contrived. Otherwise, how can dues be justified? My old Quaker grandfather, owner of a textile mill, knew this, but I had to learn it the hard way. I finally evolved a technique for dealing with the political realities of the Oil Workers Union, and it went something like this.

The local union committee and I would meet and bargain for a few weeks, sometimes far into the night. Originally this was done on neutral ground—rented hotel rooms, which Ashland had to pay for. Later I broke all precedents by offering to meet in their union hall. This saved us money, and where we met made no difference anyway. Most of the early bargaining involved minor grievances, work rules, and relatively inconsequential matters. Generally, I let the union boys win a reasonable number of minor matters, while I held firm against costly featherbedding and idiotic work rules. I usually avoided the "money issues" until the end. Money being on the line usually brought solutions to troublesome questions which had been held in abeyance. These always involved work rules, seniority, job descriptions and the always-debated problem about who had the right to run the refinery, management or the union. For these stormy sessions, I always saved my worst off-color stories. When the shouting got too loud or the tension too great, laughter helped.

An impasse was always reached at the local level. This was my signal to call Ben Schaefer, the national vice-president of the Oil Workers Union, under whose wing our local unions fell. I would usually, at some point, call him and say that, to avoid a strike, he should come talk sense to the union's bargaining committee. He would come. We would meet privately, agree upon where I could yield, what I had to have, and our strategy in yelling at each other for a few days so that he could put on a good show in front of the bargaining committee. After the proper dramatics, we came out where Ben and I had agreed to end days before. Faces were saved, and all was sweetness and light, more or less, for another year. Usually it took all-night sessions to get there.

Only once did this strategy almost backfire. After a meeting with Ben, I cleared the results with my chairman. He exploded and wanted a better deal. He demanded to address the union committee in order to explain the competitive position of the independent refiner, and I asked him to be my guest. He gave a brilliant, expert, logical, and sound exposition for some forty minutes. When he was finished, Ben Schaefer, in the presence of the bargaining committee, sunk to a profane level, which was rare for him. He turned on Blazer with a single sentence: "Paul [not *Mister* Blazer], I never heard such a lot of horseshit in my life." The committee loved it—their champion was not afraid of the big bad wolf. Paul asked for a recess and told me to try to settle the contract on the basis Ben and I had established earlier, but it took another few days of face-saving bargaining.

I discovered, not to my surprise, that union leaders were rather expert at committing "unfair practices." Once, when the Oil Workers, a CIO industrial union, were attacked by the craft unions of the AF of L, the entire maintenance crew in one of our refineries walked out. Our CIO operators kept the plant running. After a few weeks, Ben Schaefer of the Oil Workers came up with a wicked suggestion—post a notice telling the strikers to return in twenty-four hours or lose their jobs. As Ben suggested, only the leaders of the AF of L factory would not return to work, and we could get rid of the troublemakers by firing them. It turned out that way. Had an employer suggested such a course, he would have been dragged before the National Labor Relations Board in nothing flat. But, then, few employers expected impartiality from the Labor Board, and over the years this hasn't changed.

Once any new contract was signed, I had to pay my "rent" for the use of the union hall for the bargaining sessions. It generally consisted of some new piece of furniture. After that was delivered, we had a beer bust with some of the bargain sinking to just straight gin. From all this came some reasonable, personal rapport with many of our men. For me came an understanding of unions and union negotiations which could have been learned no other way. It proved highly usable in subsequent years with other companies. On a par with Ben Schaefer's "unfair practice" played on his AF of L rivals, another chairman of mine (when I was with Signal Oil and Gas) was charged with unfairly resisting an attempt by the Oil Workers to organize one of our natural gasoline plants. My chairman threatened some of our people in a speech at a company dinner. I asked what he said. He replied: "I told them they were free to join any union they pleased, they could even join the C. I. O. (Oil Workers). All I wanted was the name of the first s.o.b. who joined."

Looking For Oil Abroad: American Independent Oil Company

Before World War II, important crude oil production in foreign lands was largely owned by a very few big companies. Standard of California and the Texas Company, in the mid-1930s, were the first to challenge Jersey, Shell, Gulf, and a few others in the Middle East. Now it was time for some strong independents to take a run at the giants. In 1947, Ralph Davies put a consortium of eleven independent oil companies and individuals together named the American Independent Oil Company (Aminoil), and he asked me to have Ashland join. The original trustees included me, K. S. "Boots" Adams, president of Phillips Petroleum, and Barney Majewski, of Deep Rock Oil. Among the others, whose names are mentioned elsewhere in this book, were Will Reid of Hancock Oil, I. A. O'Shaughnessy of Globe Oil and Refining, Sam Mosher and Russ Green of Signal Oil and Gas, and W. W. Vandeveer of Allied Oil Company.

We sought a concession in the Kuwait neutral zone, just south of the great Burgan field in Kuwait proper. Burgan was then, and today still is, probably the largest oil field ever discovered. As early as 1945, DeGolyer told me—after only a dozen wells had been drilled—that he had trouble holding its "proven" reserves "down to fifteen billion barrels." The formation consisted of thousands of acres underlain by a thousand feet of highly porous and beautifully permeable, cretaceous sandstone filled with oil right down to the closing contour of an immense anticline. As De put it, on the U. S. Gulf Coast the structural jugs are a third full, if you are lucky. In the Middle East, they seem to be completely filled. We went looking for a twin, a sister, or something like it. Some optimists looking at our geophysics thought they saw something even bigger.

The story of how we happened upon the possibility of a concession in the neutral zone is almost unbelievable. One day an ex–U.S. Army sergeant, a welder from Cisco, Texas, named Jim Brooks, showed up in Phillips' head office in Bartlesville, Oklahoma. He had been stationed in Kuwait for a time, where he kept the plumbing in the ruler's harem in repair. Water was Kuwait's scarcest resource; Brooks's faucet work evidently resulted in a great deal of it being saved. To express his thanks, His Highness Shaikh Ahmad Al-Jabir as-Subah, the ruler of Kuwait from 1921 to 1950, told Jim that, if he could get some good American independent oil company's interested in the neutral zone, he would see that they got a concession. Phillips got Ralph Davies involved because of his post-Standard role as an independent and his political connections. Aminoil decided to risk a couple of air fares to send Nate Eisenberger, a Phillips geologist, and Garth Young of Signal over to Kuwait to take a look. Amazingly enough, there was some truth in the sergeant's tall tale. After a "Dutch" auction, with our agents housed in the palace and the competition looking in from the outside, Aminoil was awarded a concession on the ruler's half of the neutral zone on June 14, 1948. (The full story is told with relish by Garth Young in his tribute to Ralph Davies in *Ralph K. Davies: As We Knew Him*.) The other half belonged to the Saudis, who leased to Paul Getty.

Attending the tumultuous board meetings of Aminoil was like nothing I have ever experienced in any board of any company before or since. It never came to fisticuffs, but it was close. Before any well was even started in the neutral zone, a geophysical map was displayed, which made the great Burgan Field to the north, discovered in April, 1938, look like a small appendage on a super-giant feature squarely placed in the middle of our concession. From that moment, Phillips, the largest but by no means the controlling stockholder of Aminoil, sought to gain a majority position in the company. Some of the independents, probably in return for financial or other favors which Phillips never hesitated to use, linked up with Phillips. The remaining independent stockholders, holding a bare majority of the total stock of the company, banded together in a voting trust to resist Phillips on the general theory that they had better "hang together or they would all hang separately."

From the moment the various owners had aligned themselves as either pro- or anti-Phillips, board meetings took the form of courtroom brawls. A federal court reporter kept a verbatim transcript of everything that was said at every meeting. Directors spoke largely "for the record" in an effort to catch the other side with a legal foot off-base. Rather than concentrating on the serious business of the company,

Left to right: Garth Young of Signal, Essat Gaffar (aide to the ruler of Kuwait), and Jim Brooks of Cisco, Texas, whose plumbing skills began the quest for a concession. *Courtesy Prentice-Hall*

everything seemed to be played for future litigation between the stockholders, and this went on—literally—for years. It was a wonder that any wells ever got drilled on the giant structure thought to exist in the neutral zone.

Somehow, after a little more geophysical work, a first well was spudded in late 1948. It seemed ridiculous later, but that first well was located down the flank of the hypothetical structure to avoid the "gas cap" postulated for the top of the "high" everyone was sure was there. That first well was as dry a dry hole as was ever drilled. An optimistic geologist from Phillips claimed to smell live oil in the cuttings from the well, but Garth Young of Signal said he smelled only camel dung.

With twenty-twenty hindsight, which comes only after a series of wells have been drilled, the big "high" was there. Alas, it was too high. The prolific Burgan sands to the north had either been removed by erosion when the "high" stood above the level of an ancient sea, or the sands had never been deposited in the first place. In other words, we had a bald-headed structure. The Oklahoma City field in this country was bald-headed, but there it helped. An unsuspectedly prolific Wilcox sand pinched out against the dome to form a trap for a billion-barrel oil field.

Later, all of us had to learn again what many of us thought we already knew from this neutral zone experience. No geophysical map ever shows for sure whether an oil field is present. Oil is still only where the drill finds it. A partner of mine even said he never believed we had a good well until the check cleared the bank.

While we never found any prolific Burgan sands on our structure, we finally found another producing zone in 1954. Wafra No. 4 was not Middle Eastern in either size or quality—only medium gravity, high-sulphur oil with perhaps a billion barrels of ultimate production, most of which would have to be pumped rather than merely allowed to flow. For the Middle East, this was truly the runt of the litter.

The second-rate discovery took a lot of steam out of the ruckus to decide who was going to control Aminoil, but still the battle raged on. Phillips finally broke the log jam by persuading Paul Blazer, for considerations unknown to the rest of us, to run out on the voting trust agreement Ashland had signed. This precipitated the long-expected court battle in an unexpected form. Interestingly enough, Ashland argued that the voting trust was invalid based upon words which Paul Blazer himself had inserted in the agreement. I am sure these words were not inserted deliberately to render the instrument invalid, since Blazer had no compunction about taking advantage of the possible legal loophole. Both Blazer and Ralph Davies hated each other and fought like a couple of alley cats—perhaps because both were able, though different, eccentrics. I found myself caught in the middle. Although I was regularly lectured by Paul as a fool for having taken Ashland into Aminoil, when that company was finally sold out—yielding a substantial profit to Ashland—I don't remember any wire of thanks from him. You're often praised for some of your worst achievements and criticized for some of your best—you find consolation only in knowing the difference.

An illustration of the depth of the competitive conflicts on the board of Aminoil surfaced at one of its meetings. Some of the member companies were merely producers, interested only in the high dollar from any oil American Independent sold. Others, like Ashland and

Phillips, had extensive refining interests. They wanted to buy crude oil cheaply, and Blazer had once argued for a low price. Barney Majewski looked across the table to ask Blazer a question: "We aren't trying to make money off each other, are we, Paul?" Paul's answer, so typical of him, was, "Well, no, not necessarily." He seemed mystified by the ensuing gale of laughter. Actually the whole argument was a tempest in a teapot. The quality of the newly discovered crude was so lousy that Aminoil had to build a topping plant in the neutral zone just to get rid of its oil at any price. None of its stockholders wanted to buy much of the stuff that came out of the plant.

After Phillips got control of Aminoil by buying Ralph Davies's interest for ten million dollars, they hired a new president to replace Davies. I met him enroute on my way back from Argentina after negotiating an exploration contract with the Argentina National Oil Company. He blandly told me, a perfect stranger, how he had solved a tax problem in a similar contract he had negotiated for another company. He had retained an employee of the Argentina National Oil Company to assist him. When I asked if he paid the employee, he replied that he had and named the figure. I choked. When I got home, I learned that he was Dan Dunaway, the new president of Aminoil.

Falling Out with Blazer

Ashland director meetings were something of a dramatic production, with Chairman Blazer the star of the show. He thoroughly enjoyed explaining in great detail precisely what the company proposed to do and asking for the board's concurrence, which was a foregone conclusion. He knew his explanations served to show how able, knowledgeable, and shrewd the chairman was. As a matter of fact, he embodied all of these characteristics to a very high degree. Still, if there had been any dissenters, it would have been a case of "all those in favor say aye and those opposed submit their resignations." Still, really well-managed, aggressive companies are not run by committees, even if they are called boards of directors.

Boards are mainly useful for firing the management if things go badly for the company. They aren't an effective instrument of democracy to vote on matters about which management knows more than outside directors can ever hope to understand. Only once or twice did I witness a little humor during the course of an Ashland board meeting. When we acquired the Allied Oil Company, its former chief executive officer, W. W. Vandeveer, was elected to the Ashland board. He held many hundreds of thousands of shares of Ashland stock; thus, he felt perfectly free to express any opinion he pleased. When Paul rather seriously told his board that he had never made an important mistake during his business career, Van looked up and said: "Of course not, Paul, for whenever you did you always figured out how someone else made it." Some of us suppressed our smiles. At another time, after we had acquired the Frontier Refining Company, Blazer, who was anything but an expert in crude oil exploration, delivered a monologue on Ashland's crude oil "policies." He turned to Jim Breuil, the former head of Frontier and an able producer, to ask what he thought. Jim responded: "I never heard such a lot of horseshit in my life." Paul dropped the subject like a hot potato. Jim, like Van, owned a big block of Ashland.

On January 15, 1951, I became vice-chairman of Ashland and Rex Blazer, the nephew of Paul who had joined the company with the Allied merger two years earlier, took my old spot as president. As Otto Scott states in his history of Ashland, my new title was an ambiguous honor. This was a sign of change—and, in my case, not for the better. My relationship with Paul Blazer had changed.

By the end of 1951, my time had run out with Ashland, or, more particularly, my working and personal relationship had run out with Paul Blazer. I suspect I was the only partner Paul ever had. While we were building Ashland, it was great fun, as we both had different talents and sought to use them. Abruptly, our relationship changed from that of a partnership to one of master and servant, and I have never been able to explain why. Paul was a colossal egotist—a characteristic that made him extraordinarily good at what he did. Perhaps things would have been better if all our mutual friends had confined themselves to telling Paul what a great job he was doing with Ashland instead of about the job "you and Howard are doing." Then again, Paul was inherently suspicious of everybody and everything. He was superb at detecting crooks rolling spitballs before shooting them, but not too good at dealing with honest people. He always suspected spitballs, even when none were rolled. Paul's health was weakening—he was diagnosed with a heart condition in July, 1950, and had a major heart attack five months later. Otto Scott, in his history of Ashland and Blazer, *The Exception,* documents the strained relationship with limited explanation. All that I knew was that, in my last months at Ashland, no matter how successful my actions were, Paul always told how it should have been done differently. He was impossible to satisfy, and this fact spoke volumes.

After Christmas, I went to Paul. I reminded him of a conversation we had had when I joined Ashland. I had told him that Ashland was his company and that, if we had a serious disagreement, I would eliminate myself. That time had come. Paul said this did not fit his plans right then, and he countered my offer to leave with a proposal—that I return to the practice of law, while Ashland paid me a retainer equal to my salary "just to work on problems" he and I thought important to Ashland. I demurred, and he failed to understand my decision. I said that, if I told him, he still wouldn't understand. Nevertheless, he wanted to hear my explanation. "If you take the king's schilling, you must work for the king," I told him. He heard but missed the point. We finally agreed that I would stay long enough to finish any unfinished business. He also offered a severance allowance so inconsequential as to verge on the absurd—a little more than a month's salary—to his former partner, who had never taken a day's vacation or sick leave in seven years. People do strange things when angry.

Immediately after New Year's, 1952, a hemorrhaging ulcer laid me low. I survived the first attack, but another attack a few months later cost me a large part of my stomach. I felt as though my life had turned to sawdust. The end with Ashland came with my resignation on January 10, 1952, and this press release by the company: "J. Howard Marshall announced today that he has resigned as a Director and Vice Chairman of the Board of the Ashland Oil & Refining Company. In connection with his future plans, Mr. Marshall stated that he intended to continue his career in the petroleum industry concerning himself, as he has in the past, with both its legal and operating phases. Mr. Marshall will remain in Ashland for the next several weeks for the purpose of winding up various company matters."

This press release could not begin to explain the complicated situation or my acute disappointment over the whole affair. Right or wrong, I did believe Ashland, as it then stood, was partly my creation. Like Paul, who had hired me seven years earlier (supposedly as his successor), I was proud of it. I had presided over a company that had grown to be one of the twenty largest petroleum firms in the United

States. Refining capacity in my tenure had increased seven-fold to one-hundred and fifty thousand barrels a day; net sales likewise increased seven-fold to $150 million per annum. Most importantly, Ashland had been consistently profitable.

One of my friends claimed I had no ulcer, only a bad case of "Blazeritis." I could have answered the charge by paraphrasing Churchill's statement that he had gotten more out of alcohol than alcohol ever got out of him. I got more out of Blazer than he ever got out of me. He certainly taught me far more than I could ever teach him.

CHAPTER SIX

Producing at Signal Oil & Gas Company, 1952–60

With the disintegration of my Ashland career, it was necessary to regroup. I remembered how Ralph Davies had never really forgiven those who'd done him in at California Standard, and I firmly resolved not to fall into that trap. Hate hurts the hater more than the hated. On balance, I could never repay Ashland or Blazer for what I had learned or the opportunities afforded.

To my surprise, offers floated in. An old client and close personal friend, Sam Mosher, head of Signal Oil and Gas, journeyed to Ashland, where I was recovering from the first ulcer attack, and offered me a Signal directorship. By this time I'd helped others make a lot of money, but now it was time to try and make some for myself. I had no yen ever to be only a salary or wage slave again.

Sam provided me with an escape hatch. As part of the offer, he suggested a vice-presidency to run all of Signal's operations east of the Rockies for two-thirds of my time. The other third belonged to me to do anything I pleased for my own account in the oil industry, providing that it represented no conflict of interest with Signal. Sam and I would decide in each instance whether any potential conflict was involved. I accepted.

One bright spring morning, my family and I packed the last of our

Sam Mosher, founder and president of Signal Oil & Gas. *Courtesy Hal Littlefield*

personal belongings in the family car, waived a fond goodbye to Ashland, Kentucky, and set out for Fort Worth, Texas. I never returned for years, for, until Paul Blazer's death, I was always afraid he would hold it against my friends in the company if he heard of my talking to them. I think he thought I might try to change the management of Ashland. Although he may not have understood, the Ashland chapter of my life was irreversibly closed. Besides, in American corporate life, no stockholder revolution ever wins when a company is making money and expanding. Ashland was doing both. Whatever his shortcomings, Blazer was a superb manager in his chosen fields.

Outside of California, Fort Worth was the center of Signal's operations, which were confined almost exclusively to exploration and production. It was a relief not to have to worry, at least for the moment, about downstream markets and price wars. I once asked Max Fisher, after he sold his independent refining and marketing company, Aurora, how it felt to be part of a company like Marathon. He replied: "I'll tell you: yesterday the price of gasoline in Detroit broke six cents a gallon, and I still get a good night's sleep." In those days, Marathon, like Signal, concentrated its efforts on crude oil production—so much so that it was often said that the company didn't think a dollar was worth anything unless it came out of a hole in the ground. Signal really regarded refining and marketing as a necessary evil to get rid of crude oil production.

A Texas oil banker once described refining as "a bad habit," and Signal, together with most of the majors, shared this attitude. With a few notable exceptions, the majors paid little attention to maximizing their profits beyond the wellhead. Only the best of the independents knew how to make a good living in the oil business from manufacturing and sales. Unlike many other industries, you generally did not need to market domestic crude oil because someone would usually come and get it. Unlike refining and marketing, which must be skillfully managed twenty-four hours a day, three hundred and sixty-five days a year, once you have made a really good oil discovery, with some technical help, the operation will almost run itself. In fact, just one important discovery started each of most the modern successful oil companies, which is why such discoveries are sometimes known as "company builders." For the story of Signal's beginning in the prolific Signal Hill field near Long Beach, I'll refer you to Walker Tompkins's tale in *Little Giant of Signal Hill*.

Signal Hill, in its prime, the richest oil field per acre. *Courtesy Prentice–Hall*

Getting Busy with Exploration and Production

What Signal wanted east of California was oil production, and it was my job to help them get it. For me this was like coming home; my first love in the business was exploration and production. We started with a small and none-too-competent staff in Fort Worth, which had to be streamlined. Fortunately, we had a highly competent group on the West Coast, so we drew upon the second to supplement the first. They had to be educated on the differences between the sharp structures in the West and the more widespread features on the midcontinent. In the West, you had oil fields piled on top of each other, covering relatively few acres with hundreds of feet of oil-saturated sandstones. In the East, producing fields were spread over many thousands of acres and composed of thinner, less permeable, less porous limestones, reefs, dolomites, and sandstones. Accustomed as I had become to California production, when I was first with Ashland one of our boys called from western Kentucky and reported with glee how a wildcat well had cut twenty-five feet of pay in the Tar Spring sandstone at twenty-five hundred feet. I was so ignorant about the midcontinent that I asked whether we were going to plug the well. The field which this wildcat had found ultimately produced several million barrels and made us several million dollars. In California, unless we had several hundred feet of pay, we wondered whether a well was capable of "commercial" production.

In the 1950s, Signal built up some reasonably good, but not great, production east of the Rockies. We had a run at a few "elephants," fields which might have been "company builders," but none, alas, turned out that productive. A few times we just barely missed. Once in the Gulf of Mexico, with a group of independents, we flinched just enough at the last moment in placing our bids to lease certain offshore prospects that we lost what turned out to be four far better-than-average discoveries. On another occasion, we outbid Gulf Oil for five thousand acres on top of a big salt dome offshore. Its code name was "Jumbo." The structure was an elephant indeed, but it was dry on top with only a narrow rim of oil and gas down the flank. I think we got our costs back, but that's hardly the name of the game. One prominent independent was once presented a deal with the argument that it would get his money back in five years. His reply: "Why should I do that, I already have my money."

In exploration, no news is bad news. If your field people think they have a discovery, they will find you by telephone in the middle of darkest Africa or the far reaches of the Middle East. But if they have a dry hole, it sometimes takes days or even weeks to find out. They wait because, once in a blue moon, a well diagnosed as dry will end up as a producer. On balance, however, it is hard to make a producer out of a dry hole—if the stuff is there, it will usually come out. Even the biggest companies make mistakes. Exxon (then called Humble) drilled right through a reef, which later became the great Scurry County Oil Field, and never noticed either the reef or the oil in it. Another company, probably because of a correct interpretation of Humble's geologic information, found the field after Humble had dropped its leases. All you can ever say for certain is that if you try to follow somebody else's dry hole, or take a "farm out" from a major company, the prospect has been through one good sieve and found wanting. The story put out about the original holders being out of money or "over their budget" is generally just bull. Somehow they always find the money or change the budget to drill what they believe to be a first-class prospect.

When it came to learning the outcome of a wildcat well, Sam Mosher of Signal evolved a quicker method than waiting for a call from me. He simply asked for a "current" on Signal stock. The market always seemed to know before we did. Scouts with field glasses and portable radios find and report information fast, if money can be made. Never try to outguess the tape, some speculators say, as profitable information seems always to leak. While with Signal, I had been solicited under-the-table to buy one of our own geophysical maps. Of course, the seller didn't know it was our map and highly confidential. After that, I locked up all our geophysical data, kept one key for myself, and gave one other to our chief geophysicist with the admonition: "If anyone again ever tries to sell us one of our own maps, either you or I will have to explain it." We never did.

Great Northern Oil Company

Signal's lack of interest in refining and marketing, at least in the Mid-West, provided me with an opportunity. In my Ashland days, I had recognized the need for a modern refinery in the Twin Cities area of Minnesota to run Canadian crude and process various petroleum products then in high demand. When I had presented this idea to Paul Blazer, he'd given me a flat "no" as far as Ashland was concerned. Blazer felt a certain animosity toward Barney Majewski, a project participant, and the fact that the proposal had been mine rather than Paul's own hadn't helped matters.

My old friend J. R. Parten, who had crude to sell to the project, was still keen about the potential refinery. He asked me to join the project on a personal basis within a year after I joined Signal. After clearing the project with Sam Mosher, I joined. Now came the hard part; it would take several years and nearly fifty-five million dollars to put it all together.

These were the days of a crude oil surplus—particularly in western Canada, a frontier in a fast-moving, new oil province. First we attempted to make a long-term contract with Imperial Oil (a Standard

"Before and after" photographs of Great Northern from the *Oil & Gas Journal*, July 4, 1955.

of New Jersey affiliate) for a supply of light oil. They offered a contract, but it was based on Imperial's own posted prices for such crude. Too dangerous, we thought. Some morning Imperial's posted prices might be too high in relation to our markets for finished products in the Twin Cities to permit us to break even. At this point Imperial might well tell us to put a fire sale price on our refinery—not a pleasant thought. A fortunate alternative, however, presented itself.

Woodley Petroleum, a producing company managed and partly owned by J. R. Parten in association with Socony-Vacuum Exploration Company (now Mobil Oil), opened a series of high-sulphur, heavy-gravity oil fields in Saskatchewan. Woodley had a partner in the Southern Production Company, an American company controlled out of Birmingham, Alabama by Southern Natural Gas Company (now Sonat). Both Southern Production and Southern Natural were dominated by none other than Chris Chenery, my old opponent in the "Big Inch Oil" versus "Big Inch Gas" privatization effort—but that rivalry made no difference now. Southern, Woodley, and Mobil needed a market for millions of barrels of very poor quality crude, akin to black strap molasses. Technically, a refinery could be designed to handle it, but it would need a big catalytic cracker, reforming facilities (to improve octane numbers and furnish hydrogen for desulphurization), a sulphur recovery plant, an alkalization unit, and a delayed coker to run all the residual heavy tars from the rest of the refining operations. A delayed coker could convert these otherwise almost-worthless tars into gasoline, gas oils, and petroleum coke, which could be sold as a substitute for steam coal. At a cost of a little less than twenty million dollars, an adequate refinery could be built.

But more than a refinery was required—a sixteen-inch pipeline (the South Saskatchewan Pipe Line) had to be laid to connect the Fosterton field in Saskatchewan to the big Interprovincial Pipe Line which, in turn, crossed the U.S. border about 260 miles northwest of the proposed refinery site 12 miles south of St. Paul, Minnesota on the Mississippi river. This called for another sixteen-inch line—the Minnesota Pipe Line—to go from Clearbrook, Minnesota, to our tank farm near Hastings. The two lines covered some 856 miles in all.

Parten and I believed if we could anchor the refinery to both a firm supply of crude oil and an assured market for refined products with a fixed margin in-between, the whole complex, pipelines and refinery, could be financed without personal recourse and little equity.

Mobil, one of the producers of the heavy oil, indicated that, if we could figure out what to do, it would commit crude oil on the one hand and contract to take products on the other. Mobil agreed, as did Southern and Woodley, to price their crude oils delivered to the refinery according to the prices of similar crudes delivered to refineries in Chicago. Product markets in the Twin Cities traditionally were a little higher than in Chicago, primarily because the Twin Cities were a little further away, or at least more expensive to reach, for products originating in the Southwest.

Mobil contracted to buy some fifteen thousand barrels a day of products at a price which allowed a fixed differential, subject to escalation, over our laid-down cost of crude oil. This insulated the refinery from the risks of market fluctuations on Mobil's share of the total production of the plant. Since Mobil's credit was obviously good, their contract became bankable to finance the refinery, much the same way as Standard of New Jersey's bare-boat tanker charters supported the purchase of eight tankers by Ashland's Independent Tank Ship Company soon after the war.

Once the contracts with Mobil were made, the pipelines fell into place. These could be financed by firm contracts between the crude producers and the refinery, which, in effect, assured enough volume at proper rates to pay out the lines.

It all looked so easy after it was done. With hindsight it seemed that way to the Mobil people. Once they saw it, they went to their own management for an appropriation to do it themselves. They might have prevailed except for Brewster Jennings, the head of Mobil. J. R. and I knew Brewster from our wartime days. We were all charter members of the "Eat on Industry" (EOI) club. We simply asked our friend Brewster whether Mobil's word was good. He said it was and that was that. It was the start of a mutually beneficial relationship extending over many, many years.

When all of the contracts were executed, the First National Bank of Chicago agreed to make the necessary loans to a brand new company, and the Great Northern Oil Company was formed in 1954 to own the new refinery. Chris Chenery of Southern Production insisted that a loan participation be offered to Chase Bank in New York, Southern's principal bank. Chase almost refused when they learned the owners of Great Northern were not putting up one-third of the required capital subordinate to the bank loans. This, insisted Chase, was necessary to conform to accepted banking practice. Hugo Anderson of First of Chicago was both a banker and an oil man in his own right. He simply told Chase that, if they did not want part of the loan on the same terms as the First of Chicago was willing to accept, First would take the entire loan. This ended the objections from Chase. Of course, Great Northern had to pay a "premium" rate of interest. Unbelievable as it may seem today, that rate was 5 percent.

Since Mobil's contracts substituted for "equity," I always feared they would ask for an option to buy part of the refinery. They never did, and I can only guess they feared the lash of the antitrust laws, for certainly, had they asked, we were in no position to refuse.

One unpleasant incident marred our closing with the banks. One of our associates, Sylvester Dasone, a fairly good refiner in a technical sense who was slated to run the plant, almost threw a monkey wrench into the machinery at the last moment.

Sylvester owned one-sixth of Great Northern, as I did, and thought he was essential to supervise its construction and operation. Woodley Petroleum and Southern Production each owned one-third. First Chicago insisted that each of the owners subscribe to their pro-rata share of about $2.5 million worth of stock in Great Northern. Without warning, the night before the closing, our associate blandly told the rest of us that we would have to subscribe for his share of the stock and "carry" his interest if he was to run the refinery. I remember looking at him and asking: "Didn't we all agree that each of us was free to finance his interest as he pleased but there were to be no 'carried' interests or 'free' rides? I gave my word, didn't you give yours?" His answer: "Sure I did, but that was just for then." My answer: "You mean every time you give your word, we have to ask whether it is for real or just for now?"

We had to regroup. We got an extension from the bank. Woodley, Southern, and I agreed to pick up our associate's share of the stock in Great Northern if he failed to exercise his subscription rights. That would put my interest at 20 percent and each of theirs at 40 percent. And since no man is indispensable, we also agreed to find ourselves a new president and general manager. We followed a lesson first taught me by an experienced investment banker: "If you want to eliminate a promoter, make him come up with his money fast." We gave our associate and ourselves a short fuse—two weeks to put up or shut up.

Woodley and Southern had no problem coming up with their

share; each could write a check. Our associate, a relatively wealthy man, could have come up with his but did not. I could not then write a check for my share without it bouncing badly. I had two weeks. In that time I was able to make a deal with A. G. Becker in Chicago. We knew each other well from my Ashland days. Scraping the bottom of my cash barrel, a barrel depleted by my years of government servitude, I dug up $160,000. Becker loaned me $340,000 for five years with interest at 3 percent. The loans were to be secured by all of the Great Northern stock I had the right to buy (20 percent of the company after our associate took himself out), which included both the stock bought with Becker's loan to me and that bought with my own funds. I did not have to pledge my personal credit for the loan. This meant I could lose my own investment of $160,000 if Becker had to foreclose against the Great Northern shares but *no more*.

Naturally, Becker wanted a "kicker," a chance to make a lot more than merely interest on the money. It got it in the form of an option to buy 8 percent out of my 20 percent share of Great Northern at any time within five years at my original purchase price for such shares. Running ahead of the story, I paid off the loan within three years, and they exercised their option. Over the subsequent years, I kept buying back "my own" stock from Becker and its clients at prices which, in total, far exceeded the original loan. By then, the company was highly successful, and I only wished I had not had to borrow in the first place. Still, 12 percent of something good is a lot better than 20 percent of nothing—something our too-greedy associate failed to learn in time.

To find a new president for Great Northern, I borrowed the trading trick Paul Blazer had used on me. I visited W. J. "Bill" Carthaus, one of the top executives with Deep Rock Oil, at a meeting of the American Petroleum Institute and asked for his suggestion for a man to run Great Northern. After Parten and I described the project and the incentives, including an option for an interest in the company, he came up with the question which we hoped he would ask: "How about hiring me?" He was hired, did a great job, and really earned his interest in the company. It always seemed to me he got the most fun from a regularly renewed company car—a black Fleetwood Cadillac. Parten and I always suspected that, as a youngster making his way upward, he always wished for a big car like the big shots in the business drove.

For me, my Great Northern interest was not quite as riskless as I have described it. Our contracts with the financial institutions and Mobil made each owner agree that, if the new refinery did not perform as we had specifically represented, we would spend our own money until it did. One of my Signal people saw me pondering over the signature page of the final documents. In addition to the lines on which the banks would sign, there was a space for a signature from each representative of the highly solvent oil companies, and a line for the signature of "J. Howard Marshall, II, as an Individual." "What does that mean?" asked my Signal friend. "That means my judgment and experience better be right, or I am broke," I answered. There are few of us independents who have not had to gamble themselves or their company more than once. With the majors, probably their forbears did.

The twenty-five thousand barrel-a-day refinery came on-line in August, 1955. A twelve thousand barrel-a-day Thermofor catalytic cracker was in place, along with a 450-ton delayed coking unit. Fortunately, relatively minor start-up problems were experienced, which allowed the contractual price differentials to run their profitable course.

My happy sojourn in Fort Worth was interrupted by the unexpected death of Signal's top exploration vice-president, Donald Lycan, in October, 1952. Sam Mosher talked me into taking his place, though I never officially moved to Los Angeles, California. I kept my residence in Ft. Worth, since domestic and foreign exploration kept me out of California at least half the time. In addition, weekends were often spent in Minnesota on Great Northern matters. I still retained a theoretical right to set aside one-third of my time for my own personal business. I had trouble finding such time, except for evenings and weekends. Fortunately, that was enough.

Busy in South America

Three elephants—two foreign and one domestic—took most of my time at Signal. The first of the foreign plays was in Lake Maracaibo, in Venezuela, a country that, at the time, was the leading oil exporter in the world. Wildcat leases were offered by the Venezuelan government in early 1956. A group composed of Signal and Hancock Oil Company, both California independents, and the Pure Oil Company and

Standard Oil Company of Ohio, both (at least) semi-majors from the midcontinent, were awarded a twenty-seven thousand acre block near the center of the Lake in mid-1956. Looking at the geophysics, we all thought we had "goodie," so much so that when Garth Young (Signal's vice-president of foreign exploration), Russ Green (Signal's executive vice-president, exploration and production), and I motored over our first location in the boat, Garth opened a bottle of beer and poured it on the location with the words: "I christen thee 'Big Juicy.'"

In our imagination, it was, as Mosher said, "too good for the common people"; so good, in fact, that we turned down an offer from the owners of an offsetting block, Superior Oil Company and the Sun Oil Company, to combine their block and ours. We thought their block was "off structure." Drilling established that producing sands on their block were "lower," but a major fault separating the blocks trapped a big oil field on their block. We were left with an accumulation which got our money back plus a little. They had "Big Juicy," and we were left with only a consolation prize to show for our $38.5 million bonus bid.

Before we even had any oil to sell, the partners got into arguments like those which had divided the integrated from the nonintegrated partners in Aminoil. Both Pure and Sohio owned big refineries, which supplied refined products to their markets. Hancock had only a small refinery far removed from Venezuela, and Signal owned no refinery anywhere. Pure and Sohio insisted that each partner should have the right to take oil from a common reservoir as it pleased, regardless of the ability of the other partners to take and market a comparable quantity. Any "underlifted" oil (production under the quota, or oil that could have been produced but was not) was to be regarded as still in the ground to be made up at some future time, if ever. Mosher's opinion of that plan was simple: "I won't agree to any such scheme. Old Rockefeller played that game. He drained the leases of his competitors because he had the pipelines and the markets, and they did not. Every barrel produced from a common reservoir belongs in equal shares to each of us. Whoever lifts any one of those barrels must account to us for the rest of our share." There the matter rested. Later, when we knew we didn't have a major field after all—we were producing around thirty thousand barrels a day—the question became academic. All the oil was sold to third parties, and everybody got paid for his share.

At the same time I was in Venezuela, I kept sneaking off, incognito, to Argentina. This Signal sideline came from a conversation between Sam Mosher and Mark Millard, a senior partner with Loeb Rhoades, investment bankers in New York. Mark had already talked to me. He believed that with Juan Perón thrown out of Argentina, an oil exploration contract could be negotiated in that country. Loeb Rhoades had been retained to help Henry Holland, recently undersecretary of state in charge of Latin American affairs, and William "Billy" Reynal, a young Argentine, who belonged to the group which overthrew Perón. Reynal, the son of an Argentinian polo player and an American mother, was bilingual in Spanish and English. So too was Holland, a top flight lawyer, who had been born and raised near the Mexican border, and was a classical Spanish scholar. As Mark put it to me: "We need an oil man to figure out how these two traders should trade." Obviously, since I was employed by Signal, I could only tell him to talk to my chairman. They made a deal over the telephone. Loeb Rhoades would pay all costs and expenses. For my time, Signal could have an option, on a ground floor basis, to a 20 percent participation in any contract negotiated. Since Sam Mosher knew you never lost by taking an option, he released some of my time.

For almost two years, we journeyed back and forth between Venezuela and Argentina. Henry Holland and I stayed in a small pension, the Lancaster Hotel. None of our major competitors—Standard of Indiana, Continental, Marathon, Shell, or Jersey—ever learned we were in town until we had gotten the first olive out of the bottle. We were awarded the option that we'd requested, and only then did we move over to the big Plaza Hotel, where we knew our competitors would be staying.

To win over our major competitors, we had to invent something new and different—something more appealing to the Argentina national oil company, Yacimientos Petroliferos Fiscales (or YPF), than the time-honored signature bonus, royalty, and 50-50 split of the remaining profits, which were characteristic of the standard deal in the international oil business. Under Argentine law, no contractor could "own" any Argentine oil. This was made to order for an adaptation of the contracts made between the City of Long Beach and the LBOD. There too the city had to retain "ownership" or "title" to all of its oil until sold. It could not "lease" or assign any "title" interest in its oil to anyone. If plagiarism be the sincerest form of flattery, we proceeded to flatter ourselves. Thus, I took the form of the Long Beach contracts, which I had helped invent, changed the percentages, included a tax wrinkle, and built what later became known worldwide as the "production sharing contract."

Stripped of its legal window dressing, the contract specified that we agreed to pay the total cost of finding, developing, and producing such oil as might exist under certain defined lands. We were entitled to recover all of our costs out of a fixed percentage (60 percent) of the "value" of any oil we produced. "Value" was defined as the laid-down cost of foreign oil imported on the East Coast of Argentina. As contractors, we were entitled to 20 percent of the "value" from the date of first production as compensation for our financing and risk. This 20 percent was cost-free and tax-paid. Since YPF actually paid Argentine taxes on the "value" of our oil to another arm of the government on our behalf, we received tax receipts, which we could use to offset, dollar-for-dollar, income taxes otherwise due in the United States. Once our finding, development, and production costs had been recouped, Argentina could advertise that they were receiving 80 percent of the "value" of the oil produced, as contrasted with the typical international contract, which provided only for a 50-50 split of profits. As with so many political things, this sounded great—even if it left much unsaid. For us it offered better than the typical 50-50 deal, since, after taxes, our economics were different than theirs, proving once again that things are not always exactly what they seem.

For the only time in my life, I found myself in the delightful position where there was a chance to "wildcat" on big acreage without much geological risk. Once a decent discovery was made, all costs could be recovered both for good wells and dry holes on the contract acreage everywhere. Our first well was drilled on a geophysical structure so well-defined it seemed like we were looking at it through the cross hairs of a telescope. There was one producing field on the feature (and YPF kept it), which left no doubt about the presence of oil and only a little doubt, in our judgment, about good sand conditions to contain gas. We believed we could almost count thirty million barrels of reserves. Still, we knew better than to count before we drilled and developed.

As it turned out, we had the thirty million barrels on the first structure and, with secondary water flooding, a lot more after it. We found two more productive structures "on strike" with our first wells. All were located along the axis of a major fold, with a highway down the middle and a nearby two-track, wide-gauge railroad running eastward all the way to Buenos Aires (five hundred miles away). The last wildcat, at a location named Vacas Muertos (which means "dead cow" in Spanish), painted the way to an extremely live cow—a stratigraphic trap that none of us ever suspected. It ran for miles and contained hundreds of millions of barrels of producible oil at a relatively shallow depth. We had our "elephant" in the great Mendoza field. The last time I visited the field, with the Andes as a backdrop, I could see a line of wells disappearing over the curve of the earth's horizon. Only at Lake Maracaibo in Venezuela had I ever witnessed this phenomenon.

Often oil deals are like cats—they seem to have nine lives. "On again, off again" describes our negotiations with YPF, and we created one interesting device for breaking deadlocks. Henry Holland knew all our telephone calls were monitored, so we used that knowledge to turn the tables on the monitors. When we reached an impasse, we would call Henry's secretary in New York. If we started with her first name, "Barbara," and continued with "Tell Mr. Loeb," this meant do nothing. What she was "told" to tell Mr. Loeb related to the hopelessness of the negotiations and that, if YPF continued in its position on certain critical issues, we might as well come home. Miraculously, this stunt always got us back on track with YPF.

In addition to inventing a production-sharing formula which appealed to YPF, we devised something more: a full 50-50 partnership. All major decisions relating to exploration, development, and operations were to be made by an "Operating Committee," on which we and YPF had equal representation. One of our representatives would serve as chairman, with the right to break only tie votes until we got our money back. Thereafter, a YPF representative would become chairman. This was so unorthodox that some of the majors later charged us with "unfair competition." One of them bitterly accused me of offering something I knew they could not meet because of its effect on their operations in other parts of the world. Another told me he had been warned that I had been brought up in a "pool hall"—meaning I was an alumnus of Standard of California and knew where some of the "bodies were buried." Whatever the charges, our partnership worked. Neither of the partners had to break a tie. We debated each problem until we agreed, and, because ours was a real partnership with YPF, we survived when the political party in power changed. Others, who had a different relationship, got taken over.

It's worth a brief mention here that Henry Holland was a real professional when it came to Latin American politics. He and I agreed at the very beginning that we would either win a contract on top of the table or come away empty-handed. At various times, we were approached by various factions of the military, and others, claiming "influences." For a hundred (or several hundred) thousand dollars, we were told, a satisfactory contract could be "arranged." If the proposal came from anyone who appeared to be influential, Henry would ask for an appointment with Arturo Frondizi, the president of Argentina from 1958 to 1962. We asked a simple question: "Mr. President, is this something you want done?" Henry knew that, even if there was something behind a questionable proposal, no head of state could afford to have improper influences brought to bear if they knew about it. We always got a negative answer to our question from President Frondizi. That would be the last time we would hear about the matter—at least until the next influence-peddler came along. What transpired behind the scenes, we never knew or sought to learn.

When President Frondizi lost out a few years later, a military government led by Jose Maria Guido sought to "expropriate" our investment. I went to Sen. Hubert Humphrey, then the majority leader in the Senate and head of the Foreign Relations Committee, to explain our predicament. Humphrey asked me two critical questions: "Did you write a fair contract, and did you pay off anybody to get it?" I could answer truthfully: "It was, in the international oil business, the best contract for a host country offered up to that time, and we never paid a penny to anyone." I added that we had been importuned to purchase influence, but that Henry Holland had handled such invitations as I have described above.

After months of off-and-on negotiations, we finally reached an agreement on the provisions and the language of a contract between YPF and a corporation named LR (Loeb Rhoades) Development. As we approached the signing day, we sent for Henry Loeb, the senior partner of Loeb Rhoades in New York, for an authorized signature. As in most situations like ours, there were the usual last minute loose ends which had to be tied up. While Henry Holland, Billy Reynal, and I sought to work these matters out, we had to leave Henry Loeb in hiding in our little Lancaster pension, telling him that he was our "secret weapon." This left us in the same position as our opponents. We, like they, had to consult with our principal—our "missing man"—before we could agree to any concrete changes in the contract. One of the last points to be settled later proved to be one of the most critical. It involved a guarantee by the Argentina Central Bank of the government's performance under our contract and payment in U.S. dollars of the monies owed us. Without these guarantees, I doubt we could have survived over the years. Central banks everywhere are leery of a default, which might undermine their credit standing with other banks.

Once the contracts were executed, those of us who had carried on the negotiations for so long headed home. As a memento, John Loeb presented each of us with a pair of gold cuff links, which he had made for us at Cartiers in New York. They were engraved with my initials on one side and "Victory at Buenos Aires" and the date of the official contract signing on the other. Most of us are at least a little superstitious, and somehow, I think, these links always seemed to bring me good luck. For many years I wore them whenever I found myself involved in some important deal.

Upon my return to Signal, Sam Mosher exercised his ground floor option to take a 20 percent participation in the Argentina contracts. Some of my associates worried whether we'd bitten off more than we could chew. One, in particular, who'd dealt with our contracts in Venezuela, wanted no part of Argentina. He said he preferred Venezuela, where they had a "seasoned petroleum law." Sam put that argument to rest by asking him how "seasoned" he thought the laws were here at home. We almost lost our Tideland oil properties in 1947, when the U.S. Supreme Court reversed decades of decisions to hold that the federal government had "paramount rights" in state tidelands, over which the states had exercised jurisdiction for a hundred years. It took an Act of Congress, specifically the Submerged Lands Act of 1953, to nullify the Supreme Court decision.

Dealing from "fright" never seems to win—at least not in oil. Just a little fright cost Signal 2.5 percent of its 20 percent participation in the Argentina contract. In an effort to pacify our restless associate, who feared for our future in Argentina, Sam indicated that, perhaps, Signal should take only 17.5 percent instead of the full 20 percent to which we were entitled. I suggested we take the whole 20 percent and allow me to sell 2.5 percent for a profit, and this we did. I placed the

2.5 percent with Max Fisher, president of the privately-held Aurora gasoline company in Detroit. We allocated .5 percent to each of his three top executives, Fisher took .5 percent for himself and assigned the remaining .5 percent to Aurora. Such percentages may seem small, but in the oil industry small percentages can be large in terms of dollars. Sometime after I left Signal and Aurora was merged into a major company, that company regarded .5 percent as too small to bother with. Max Fisher asked a friend of ours to try to find a buyer. My friend came to me, and we bought it together. Over a quarter century it has repaid the purchase price many times over and still continues to do so. During that same period, the "seasoned petroleum law" of Venezuela went by the boards, and foreign oil companies in Venezuela lost most of their investments. Ours survived.

Some of us learned a simple lesson about petroleum politics. In the final analysis, "sovereign" governments, including our own, can and do what they please, regardless of their contracts or agreements. Sometimes contracts are either abrogated or ignored, new or different taxes are used to render an agreement worthless, or courts are persuaded to find technical flaws or new interpretations of old laws or constitutions. In the case of the Tideland titles in this country, our Supreme Court was "persuaded" that the Tidelands should be held for the benefit of "all of the people" rather than just for those residing in the states which fronted the oceans.

As so often happens with those who seek to do "good" without understanding underlying forces, the idealism implicit in the phrase "all the people" was only skin deep. Groups of greedy lawyers and politicians, seeking to gain cheap titles to oil fields which others had discovered and developed, also stood to benefit. They included, among others, a senior U.S. Senator, Burton Wheeler, and a former prominent New Deal lawyer, Thomas Corcoran. Devious oil men of doubtful morals were also involved. They simply plastered all known Tideland oil fields with "script" pieces of paper, issued long ago to soldiers as a reward for their service, entitling the holder to claim a piece of the "public domain" at no cost. If the Court ruled the Tidelands were and had always been part of the "public domain," the others believed such "script" would allow them to seize offshore oil fields for nothing but the nominal cost of "script" paid to the descendants of the original holders, who had never exercised their claims. It almost worked. That former New Deal attorney tried to get Signal to pay substantial sums to some of his clients, who had filed script and claimed ownership of all our state leases on the offshore segment of the Huntington Beach oil field in California. Had their plan succeeded, they, not "all the people," would have been the beneficiaries of a doubtful Supreme Court decision granted on an idealistic but empty principle.

Busy in the Los Angeles Basin

The search for oil by Signal included the Los Angeles Basin, even within a few miles of our head office. It was remarked that our own drilling people had forgotten how to drill a "straight" hole. Ever since Signal had mastered directional drilling under the Pacific Ocean at Huntington Beach, we regarded such drilling as our own particular vocation. Redondo Beach, which fronted the ocean, provided another such opportunity. After a series of political maneuvers with the city council, contract negotiations, and a victorious referendum on November 15, 1955, we earned the rights to try and find a new field or extend an old oil field discovered in 1922 in Redondo. We succeeded, but it was no elephant. We drilled a few dozen directional wells from a few selected locations, beginning in 1956. They were only moderate producers, perhaps because the field was trapped by "poker chip" shale (made up of slices of white shale which broke apart like chips in cores taken from the formation). Peak production never reached three thousand barrels a day, under 5 percent of Signal's worldwide total.

Our next "directional" play, not jinxed by any "poker chip" shale, proved to be a little elephant—at least for old Los Angeles that was thought to be largely developed. Smack in the middle of a movie studio, a starlet's house, two swanky country clubs, and the site of the original Los Angeles Open golf tournament, there arose a prospective oil-bearing structure. It had never been drilled. Literally thousands of town-lot leases had to be assembled, and a central drilling site had to be leased, from which a host of directional wells could be drilled in the event of a discovery. The golf course of the Hillcrest Country Club seemed ideal, but it was under lease to one of our competitors. Fortunately for us, the lease was due to expire in a few months. We did what was once regarded as not quite "cricket." We "top leased"—that is, we executed a lease to take effect once the other lease expired. We worked with Willard Isaacs, chairman of the club's oil committee, and, incidentally, the head of Max Factor. Hillcrest boasted more millionaires than any other country club in the country, and Willard Isaacs was one of them.

Willard was persuaded when we explained how the members of the club could benefit the most from an oil discovery. We had done it before—when we leased the lands of the Bolsa Chica Gun Club—by distributing a fractional interest in the property to be leased to each member of the club. Thus, each member could claim a depletion allowance on his income tax return, and the proceeds of the lease would accrue to each individual member rather than to the club as a whole. An added incentive took the form of a profit-sharing net royalty rather than that of only a fixed percentage. Our competition, who thought he could renew his existing lease when it expired, never knew what hit him until it was all over, and his defeat gave me great personal satisfaction. Gene Starr—the competition—was one of those who had fought every attempt to moderate the law of capture in California. There was little love lost between us.

The Hillcrest lease was the beginning, not the end, of our problems. The assembly of the thousands of town-lot leases surrounding Hillcrest, covering some fifteen hundred acres, took months. Then the Los Angeles City Council had to rezone the area to allow drilling. We submitted a plan to confine our drilling to two single soundproofed locations disguised behind imitation office buildings on two neighboring golf courses. We won in the council, but the mayor vetoed, and it took four out of seven council votes to override.

The first well in 1959 provided five hundred feet of saturated oil and gas sands. We developed the field with scores of wells—all whipstocked from one location on Hillcrest's golf course and one location on the nearby Rancho Park municipal course. To help defray the costs of the project, I invited my good friend Charlie Jones to bring Richfield in as a partner. People hardly knew that an oil operation was being conducted. Wellheads were buried below ground level, and bottom hole pumps were used to lift the oil after the natural flow ceased. A small natural gas liquids plant was hidden to extract natural gas liquids from the gas. As with Redondo Beach several years before, the whole operation—producing some forty-five hundred barrels a day—served as a perfect illustration of how neither exploration nor production need to be dangerous, dirty, or offensive if properly engineered and managed.

One fringe benefit came from our discovery. Hillcrest Country

Club served some of the best food in Los Angeles. Whenever I appeared to talk with them about operations, I was always wined and dined as though I'd somehow generated the hydrocarbons under their club. I think Hillcrest took particular pleasure in their having oil and gas and the adjoining Los Angeles Country Club having nothing but dry holes. I enjoyed watching wealthy Hillcrest members trying to trade their royalty interests with each other. When they sought my ideas of value, I was smart enough to have no opinion.

After Hillcrest came another bit town-lot play, which we nicknamed "Sugar Hill." In early 1957, my top land man in Signal, Jim Wootan, got wind of some quiet teasing by Union Oil of California concerning an area a few miles from our own building located on Wilshire Boulevard. By teasing us, Union Oil inadvertently tipped their hand. We jumped in, and before Union knew they had a competitor, we had hundreds of special lease forms printed, circulated, and signed in their area of interest. To beat Union's 16.67 percent royalty, as well as a smaller royalty offered by Guiberson & Burke, we offered 20 percent. We ended up with 40 percent of the prospective field—not bad considering our staggered start and the tens of thousands of town-lot leases involved. Then we pooled our leases with Union and sought another zoning variance. This was not too difficult, especially since "Sugar Hill" was much less affluent than Hillcrest and Beverly Hills. Here, residents wanted development and oil income.

Geologically, Sugar Hill differed from Hillcrest. Little or no surface indications pointed to an oil field. Geophysical evidence showed a deeply buried ancient hill and complicated fault pattern with the possibility of hydrocarbon accumulations trapped on or against these features. As it turned out, a trap was there—not exactly as we'd visualized, but near enough. As had happened before, a series of wells drilled directionally from a couple of locations finally pinned down the exact boundaries of a new Los Angeles Basin field in 1959. Strangely enough, it extended almost under Alex Perino's old restaurant, where we laid the dinner check in the lap of the president of Standard of California some twenty years before.

Argentina Again

While all my domestic exploration efforts for Signal were proceeding, Argentina still demanded attention, my own private interest in the

Great Northern Refinery in the Twin Cities could not be neglected, a carelessly bought Signal refinery on the Gulf Coast had to be rescued, a participation opportunity in the Iranian consortium needed evaluation, a motion picture venture for which some of Signal's foreign crude oil had been foolishly exchanged had to be handled, an important merger with the Hancock Oil Company required negotiation, and the acquisition of the Norwalk Oil Company needed to be digested. These and other problems landed in my lap. Such was the life of a newly promoted executive vice-president of Signal.

The Argentina contracts had been signed, and a solid operation was now critical. Frank Morgan, our geological consultant on Argentina, told me that in Richfield they had a man who could manage anything, and he was nicknamed "Old Jess." Whenever they needed someone to take on a strange new task, it went to Old Jess. We found our Old Jess for Loeb Rhodes Development in the person of Bruce Clarity. We baited his hook with a good stock option, and off he went to Buenos Aires.

He and I now had the Argentine task. After checking the cheapest ways to commence, we entered into a firm drilling contract that allowed Kerr-McGee Corporation to finance the restoration of half a dozen second-hand drilling rigs which Kerr-McGee owned in Oklahoma. Kerr-McGee wanted new and more modern rigs for their operations in the United States, but the old "Natural 75's," a workhorse rotary rig of its day, were more than good enough for our purposes. We put them in first-class condition with enough aluminum paint to make them look like new. Once they arrived in Mendoza, they drilled scores of wells with only a few days inactivity between jobs, compared with the YPF rigs that could sit idle for several months. Our payrolls were exclusively Argentinian except for one American "tool pusher," the boss, on each rig. All he had to do was to train his Argentine crews on how to drill quickly, efficiently, and cost effectively. Our Argentine crews were first-rate and fast learners. In a year or so, we could send the American "tool pushers" home.

Argentina, not unlike own country, provided a perfect example of the inability of governments and politicians to run anything efficiently. One day Bruce Clarity stood watching a YPF drilling crew pulling drill pipe out of a well to change the bit, an operation called "making a trip" in which the pipe is removed in sections. I saw Bruce looking at his watch. He turned and noted that the crew took almost three minutes to pull one section, uncouple it, and come back for the next section. He remarked that the whole operation should have taken less than half a minute, and, after we had retrained our Argentine crews, that was all the time they needed. We once calculated that YPF had one employee for each daily barrel of oil production. In private industry this would have been regarded as featherbedding gone wild. After we'd developed some forty thousand barrels a day of production, we had only a few hundred employees. The difference arose not from a difference between American and Argentine personnel, since, as stated above, almost all our people were Argentine. It was the difference between a private, profit-seeking company and a bureaucracy which never had to worry about the "bottom line." As with our own Department of Energy, bureaucracies spend most of their efforts administering rules and regulation, seldom do they "produce" very much of anything.

Absorbing Eastern States

Returning from one of my trips to Buenos Aires, I found that Frank "Dutch" Lortscher, one of our Signal vice-presidents, had negotiated the purchase of the Eastern States Petroleum & Chemical Corporation, located just outside of Houston. It was controlled by one of my old hot oil opponents, Dick Kahle, who was notorious for his ability to consume large quantities of alcohol and still do business. The deal for Eastern States was so bad for Signal, I could only conclude that Dick had held his liquor far better than our representative. Dick was given Signal stock for his company without even being asked to warrant his own balance sheet. No one in Signal except myself knew much about refining in general (particularly on the Gulf Coast), and what I knew came from my experience as president of Ashland.

Not knowing what else to do, Sam Mosher dumped the whole Eastern States affair in my lap. First, I learned that someone had "neglected" to show on Eastern's balance sheet millions of dollars of liabilities for tankers that had been chartered for long terms at high prices during the Suez crisis. To settle these claims, I almost had to go back to the practice of law. It took several years, but with the assistance of an old friend of mine, Dave Scoal (who was a top flight maritime lawyer), we fought, bled, and almost died in the court and arbitration proceedings. We escaped, however, at a cost of only a couple of million dollars plus the inevitable legal fees. The charters were with various Greek shipping magnates not known for their charitable instincts. We were fortunate to escape as well as we did.

Next came the condition of the refinery, an old wartime cat cracker located on the Houston Ship Channel. I found a far better, modern cat in good condition standing idle at Destrehan on the Mississippi River. It was part of a closed refinery owned by a subsidiary of Standard of Indiana, and the equipment could be bought at junk prices. It seems ridiculous today that a twenty-thousand barrel, first-class cat with valuable storage facilities and boilers could be purchased for about two hundred thousand dollars. Working with Brown & Root, one of the best refinery construction outfits on the Gulf, I found that the Destrahan facilities could be bought for this figure, "as is where is." After paying for the costs of removal, transportation by water to Eastern States, and reerection, we could have ourselves a modern "Model Four" fluid cracker ready to run for about $1.5 million. At that time, a brand new unit of equivalent (but no better) capacity would have cost many times that.

I requested authority from the Signal board to purchase and reerect this unit before the old one at Eastern States (both economically and literally) blew up in our faces. I explained, using the terminology of Max Fisher, how a cat cracker was the cash register of a refinery because it converts low-grade, low-priced heavy materials into high-quality, higher-priced, light hydrocarbons—gasolines and heating oils, for example.

As a result of mergers and acquisitions, Signal had a twenty-six man board, and, to my astonishment, the board refused my request for funds. I was convinced we would regret such a decision, and, after the board meeting, I told our chairman in no uncertain terms that, when it came to refining matters, our board had no expertise. Sam Mosher somehow sensed the board was wrong, but, even with all his personal power, he hated flatly to reverse "his" board. He pointed out that, as the executive vice-president of Signal, I had the authority to spend up to five hundred thousand without board approval. Could I split up the purchase of the cat into separate contracts—so much to buy the unit, so much to transport it, so much to build new foundations at Eastern States, and so much more to reerect the cat? Knowing full well that Sam's vote was the only one that mattered, my reply to his question was simple: "That is easy, Sam, if you tell me to do it." He did.

The fragmented contracts were made, each for less than five hundred thousand dollars. When all was completed and the unit was operating, Russ Green, who had been president of Signal since the beginning of 1958, came storming into my office. He accused me of deliberately running around the decision of the board. I confessed but suggested we talk to Sam, who settled the matter by telling Russ he'd told me to do what had been done. Some years after I'd left Signal, President Russ ran into me at an API meeting. He just wanted to tell me how thankful he was that a way had been found to buy the cat for Eastern States. Said Russ: "It saved our bacon many times."

Another potential source of red ink for Eastern States came from an almost complete lack of captive markets for either its gasolines or fuel oils. So serious was this deficiency, that Dick Kahle, our products marketer, sometimes had to start tankers from the refinery for points north of Cape Hatteras without a buyer in sight. In such circumstances, poor Dick was easy prey for every oil broker on the north states' coast. Whenever any refinery must free up tank space by shipping without a known buyer, the refiner is asking for trouble. I struggled for months to get Eastern States out of this trap. Using both stock and cash, we acquired a large independent fuel oil marketer, with captive business in New York and New England, and two substantial independent retailers of gasoline in Florida and the midcontinent. Both owned, leased, and operated a chain of "cut price," high-volume, big dump service stations.

We now had a reasonably good refinery, auxiliary equipment to produce aromatics (like toluene, xylene, and benzene, which are often loosely referred to as petrochemicals), and captive markets for conventional refined petroleum products. But one important piece was lacking—assured supplies of crude oil at the right price. Fortunately, Eastern States, unlike many refineries, was equipped to handle West Texas sour crudes. Crude oils in California at that time were relatively less plentiful, in terms of current demand, than those in West Texas. This facilitated exchanges between Signal's own California oil production and refiners in California having excess oil in West Texas. In addition, foreign sour crudes were available, such as the oil produced by Aminoil in the Middle East.

One other important refining facility was added. Benzene was in very short supply, and my old company, Ashland, had perfected a process for converting toluene into benzene. To convert our own toluene, as well as additional amounts purchased from other refiners, we built a hydrodealkalization unit under license from Ashland. Before we started construction, we entered into a one-year contract to supply one thousand barrels a day of benzene at a price high enough to pay for the entire cost of the new unit, plus a profit, within the contract period. For the first, and I think the only time in my life, a new unit started up on schedule, without a hitch, within budget, and in excess of its design capacity. Had it not done so, we would have sold benzene short with no way to cover except at a prohibitive cost. I slept better when the new facility came on-line. Short sales are dangerous in markets other than those reported on stock exchanges.

Managing Great Northern

While Eastern States's refinery problems were in reasonable order, some of my own time was spent solving my own problems with the Great Northern plant in Minnesota. Here, the plant did not start either on time or without hitches. A picket line unexpectedly showed up to block the entrances to the plant while we were hiring operators to commence operations. After a week or so, I discovered that the picketers were members of the Teamsters Union, so there was nothing to do but try to see Jimmy Hoffa, the leader of the Union. None too politely, he demanded that all our plant's prospective employees be required to join the Teamsters. After we overcame this initial hurdle, we found bugs in the design of the plant, which took some months to cure.

All new enterprises have growing pains, and the Great Northern refinery near St. Paul was no exception. Its president, Bill Carthaus, was clearly one of the best marketers and traders of refined petroleum products in the business and was respected and admired by majors and independents alike. But he was not the greatest when it came to managing a refinery operation. Few men excel at everything. To help Carthaus, we first tried an able refinery engineer to run the plant. Unfortunately the union tried—and almost succeeded in—running *him*. Moreover, he proved unable to control his capital expenditures; I had to approve every expenditure over one thousand dollars.

Then I found a man suggested to me by my old friend, Max Fisher. Art Winter had run an independent operation in Michigan but had left because he and his wife, the owner's daughter, had parted. He'd since remarried and gone into the machine tool business in Arizona, but now he wanted back in the oil industry. J. R. Parten and I hired Art as a future top executive of Great Northern. Although we made it plain to him what we wanted, we had to, ultimately, "smuggle" him into the job. Initially we made him the controller, which gave him a year to familiarize himself with operational details. Then we elevated the president, Carthaus, to chairman and made Art our president. In this arrangement Winter headed the physical operations and Carthaus continued to direct overall policies, sales, and negotiations with other oil companies.

When, like Signal with Eastern States, we needed some captive markets to assure capacity operations, the chairman made two "impossible" acquisitions. One involved about half of a high-volume, cut-price chain of service stations spread throughout the Middle West. The other included all of a large, independent terminal and fuel oil marketer distributing in Chicago and on the Great Lakes. Almost none of the gasoline or fuel oil outlets of these two acquisitions were located near our refinery in St. Paul. Yet, since we had the best and largest refinery in the St. Paul area, and others marketing in this region had to bring in their supplies from distant sources, exchanges could be arranged whereby we could furnish their supplies in areas naturally tributary to our plant. They, in turn, could supply our remote markets from their tributary refineries. Some may argue that "exchanges" foster monopoly, but what we accomplished intensified competition over a far wider area than could be reached by actually moving the product from a refinery located at the head of navigation on the Mississippi River.

Some of my Ashland lessons had to be applied, sometimes none too gently, to Great Northern. Our plant was designed to run a little over twenty-three thousand barrels a day of low-gravity, high-sulphur Canadian crude oil with no output of heavy residual fuel oils, unless market prices justified such production. After we got the bugs out of the new plant, we started out at twenty-six thousand barrels a day. Month-by-month that number increased up until reaching forty thousand. We experienced only minor capital expenditures to remove bottlenecks in the plant after learning about them through trial and error. Again and again, I was told, we had reached the limit; again and again I had to order the throughput to be raised by a few hundred barrels. Only this way can maximum capacities be determined. Once,

J. R. Parten and I had to threaten our very able chairman with possibly losing his position if he failed to continue to move up our throughput. It went up, nothing exploded, and our profits went up proportionally.

Several Acquisitions

Now it was back to problems with Signal, which were somewhat different than the challenges posed by Great Northern. In the late 1920s two new independents had fought their way into the highly competitive California oil industry. Signal had started with a small natural gasoline plant to remove liquids—mostly natural or casing-head gasoline—from a surplus of gas produced with flush oil in the great Signal Hill field near Long Beach. The plant was built from a Bureau of Mines pamphlet with family money, despite the fact that Sam's father had "no confidence in [his] business judgment." The other independent with a future was the Hancock Oil Company. Hancock built a small, conventional skimming plant on Signal Hill to top cheap surplus oil thrown on the market by town-lot drilling and the law of capture. In succeeding years, both Signal and Hancock got involved with oil production. Both frequently joined together—for example, in the development of oil fields at Huntington Beach and Wilmington—and both battled each other (and others) tooth and nail up and down the West Coast marketplace.

The different growth rates of the two companies reflected a difference in the philosophies of their respective chief executives. Will Reid, head of Hancock, was a financial conservative. He maintained an impregnable cash position, waiting for the next "depression" for bargain purchases. Sam Mosher, however, was not content to wait. He expanded with every cent that Signal could, as the saying goes, "beg, borrow or steal." I once remarked that the worse our balance sheet looked, the better we were doing because we were using, profitably, every dollar we could muster. Leverage is the best game, provided you can make a good profit and keep the creditors off your back.

Since there was no repetition of the crisis of 1929, Sam Mosher's economics proved better than Will Reid's. By the middle 1950s, while both companies were successful, Signal was perhaps twice as large. Will Reid, who had always resisted a merger with Signal, died in April, 1956. The original drive which had built Hancock seemed lacking, while Signal was still driving. These circumstances led to a merger of Hancock into Signal, with Hancock stockholders taking Signal stock for their shares effective October 30, 1958. As usual with most mergers, and despite promises and protestations that nothing much would change regarding the personnel of the acquired company, in a year or so most of the old Hancock executives were gone—either by death or resignation—and only the more venturesome Signal group remained.

In American corporate life, this is generally what happens—for good reasons or bad, only those from the surviving corporation survive. Momentarily, the surviving corporation may get a few new directors, but they do not stay for long either. When Union Oil of California acquired Pure Oil by merger, even the former president of Pure did not last long, not even as a director of the surviving company. The story was no different when Atlantic-Richfield (ARCO) acquired Sinclair Oil. The same fate met my good friend Art Winter, president of Great Northern, when that company was merged into Koch Industries, Inc. It would happen to me, although not as quickly, after I merged Union Texas Petroleum into Allied Chemical in 1962, which is the subject of the next chapter.

One facility acquired in the Hancock merger was their Signal Hill refinery, which had been expanded and improved over the years. It was a reasonably good twenty-one thousand barrel-a-day plant with a three thousand-a-day cat cracker and other equipment for improving and increasing the quantity and quality of the more valuable, extractable products from crude oil. But it was crowded into a small parcel of land that was more valuable as real estate than as a refinery site. So crowded were the facilities, that a small fire almost destroyed the whole refinery. Signal really didn't need it to convert crude oil into products, so we shut it down and collected insurance. Other assets from the merger included several crude oil pipelines, a marine terminal, several tank farms, production in five foreign countries and nine states, a chemical plant, and service stations. We also got Will Reid's twenty million dollar cash horde.

About the same time, young Russ Green, Jr., an able son of Russ Sr., Signal's president, came up with an idea for another merger. Russ Jr. and I went to work on another old-line California independent, the Bankline Oil Company, which, in 1950, had purchased the Norwalk Oil Company. Bankline, located near Bakersfield, owned a chain of branded, cut-price gasoline stations and a refinery, which housed the equipment for converting heavy San Joaquin Valley oils into light products. Its assets looked like a good fit with Signal's. In terms of its people, however, time had run out. They were headed by an older oilman, named Red Aubert, a friend of mine from the days of the Central Committee of California Oil Producers. Once, these companies had been the private preserve of Al Weil and his client, an old steamship master named Capt. H. J. Barneson, who founded a much larger company (which later became General Petroleum, Mobil's West Coast subsidiary). As with most of those who are ready to cash their chips in and go fishing, a deal was not difficult. After the normal arguments about what their stock was "really" worth and the "true" value of ours, we hit upon the "right" ratio that represented a $6 million value. They took our stock for theirs on July 10, 1959. Signal, for the first time in its thirty-seven year history, was a fully-integrated oil company.

The ten thousand barrel-a-day refinery at Bakersfield was a strange thing—it had a fluid coker but no cat cracker. The coking unit could convert residual tar into light material and coke. But much of the light stuff from the coker and the atmospheric distillation units needed a catalytic cracker and its auxiliaries to maximize the yields of high-quality gasolines. Signal had all the equipment required to balance this Bakersfield refinery, which was sitting idle and had been only slightly damaged by the fire at Hancock's old refinery at Signal Hill. We tore down, moved, and reerected that equipment, though refinery managers always claim it's better to build new than to worry about second-hand equipment. Better it may be, but whether it's more economical is another story. Ashland built an important refining empire with low capital costs per barrel by using complete refineries and reerected units which had been used for years, and I followed Ashland's experience. It took an order to get it done, but it saved and made us a bundle of money.

Gasoline Marketing at Signal

The acquired companies marketed gasoline up and down the San Joaquin Valley, an area where Signal's marketing efforts were weak. This was part of the "fit." Signal, even without a refinery, was a successful marketer without any subsidies from our upstream operations. Marketing was run by an eccentric, O. W. "Bud" March. His was a strange operation. Every day in the late afternoon, Bud and his immediate staff, including secretaries, would gather in his first floor office, and get slightly plastered. Perhaps that was what it took in those days to understand gasoline marketing and the market. This office drinking drove Pres. Russ Green up the wall, but there was nothing he could do. The chairman, who loved his marketers, often joined their late afternoon "meetings" as a full participating member.

Personally, I got a lot out of watching Signal's marketing stunts. One group of stations sold everything from garden hoses to garbage cans. The driveways were so cluttered with merchandise, it was difficult to get to the pumps. A whole series of different brands were used by wholly owned, partly owned, or completely unaffiliated retail marketing companies or individuals. We even lost the right to sell our motor fuels under our own name. When we sold our "Signal" outlets to Standard of California, our name went with them. Bud March could not have cared less. When it was once suggested that all of our gasoline should be marketed under a single brand, Bud said no, and his reasoning was right. If you attract the attention of the major branded marketers, they must meet your prices on the nose, but if you play it low-key, they can afford to ignore you.

This was the nature of competition in gasoline markets throughout the nation in the days of almost universal surpluses. The situation was reversed only with federal regulation in the 1970s, when an independent or major with supply could sit in an office while buyers beat the doors down to buy—if the monopolistic regulations of the Department of Energy allowed the sale, that is. Allocating or dividing up the market—something that some of us may have aspired to do but never could do—never became a reality until government, the alleged guardian of competition, stepped in to deliver a fixed share of the market to "deserving" parties. With deregulation in early 1981, it was not surprising to see the "old" gasoline marketing begin to reemerge, but gasoline marketing to this day has never quite matched the intense rivalry of the 1950s and 1960s.

Leaving Signal

Unbeknownst to me at the time, my life was about to change once more. At various times since having left Ashland, offers of other positions had floated by. Husky Oil, Frontier Refining, Petrofina, and others took a run at me. Petrofina, at the suggestion of my old boss,

Ralph Davies, and Don McDonald, a partner in Blythe and Company (a New York stock exchange firm), got serious. The chairman of Petrofina persuaded me to journey to Brussels with MacDonald to talk about my heading an American subsidiary to be formed by Fina to break into the U.S. oil business. The offer was tempting, but it meant reporting to foreign individuals rather than to my lifelong friend, Sam Mosher. I declined their offer, although I did serve for a short time, with Mosher's consent, on the board of their newly-formed American subsidiary.

But now came something different. A company called Union Oil and Gas Corporation of Louisiana (to distinguish it from Union Oil of California) was owned in about equal shares by the public, the Burden family (the heirs to the old Frasch sulphur fortune) and clients of Loeb Rhoades, the New York investment banking firm with whom I helped put together the Argentina contracts. I was approached by Mark Millard, a senior partner of Loeb Rhoades, who had persuaded Sam Mosher and me to attempt to get that first olive out of the Argentina bottle after Perón was thrown out.

Millard told me they were looking for a new president of Union of Louisiana to replace a man about to retire. Would I take on the job? Though I answered in the negative, he did not allow my quick "no" to rest. He enlisted the help of R. McLean Stewart, head of Austral Oil, a former investment banker, and the director of Union of Louisiana representing the Burden interests. We met a few times for dinner and enjoyed the high living of the famous 21 Club in New York. Perhaps it was the food, perhaps the vintage wines, or perhaps just the congenial conversation with two highly intelligent oil entrepreneurs, but whatever the reasons, I listened. First, they proposed to double my salary. Then they offered a 2 percent overriding interest in any oil and gas the company discovered while I served as president. The then-president of the company had had such an arrangement as well.

The override was easy to turn down—not because it was unattractive but because an override would come "off the top," regardless of whether the discovery made the stockholders a profit. It was akin to the fraudulently labeled "windfall profits" tax from 1980 to 1988, which levied an "excise" tax, regardless of profit, on each barrel of oil output. As I told Mark and Mac, an override could possibly put me in a serious conflict of interest by personally profiting at the expense of the company. I was only half serious when I told them I might consider a big stock option at the current market price of the stock. I could make money from such an option only if all of the stockholders made more. They jumped on the idea, which was neither new nor novel to either of them I'm sure. When their bidding reached fifty thousand shares, together with a big salary, the usual fringe benefits, and assurances of full authority to run the company, somewhat reluctantly I told them I couldn't turn their offer down. I also gave my word that I would not "shop" their offer. We had reached an agreement, and no matter how Signal might counter, my word was on the line.

Now, my problem was how to tell Sam Mosher. I invited him to have dinner at Perino's out on Wilshire Boulevard in Los Angeles, one of his and my favorite restaurants. After a few martinis, I screwed up enough courage, and I told him I couldn't resist the challenge of the chief executive officer's position in a company almost as large and almost as diversified in the oil business as Signal. He responded by telling me that, as executive vice-president and director, I was marked to be Signal's top man when he and Russ Green either died or retired. "Yes," I said, "but I think it would be more fun to be captain of a ship now than serve as a first mate for an unknown period." I recalled to him a story we both heard during the war about the GI's wife who wired her husband in Australia. "What have the girls in Australia got that I haven't got?" The reply: "Nothing, but they have it here." Sam laughed and said he understood. It was the last time we mentioned my leaving until December 31, 1960, the day I was collecting my papers in the old office once occupied by Sam's father. Sam spoke briefly: "Young fellow, you have done a great job and made us a lot of money. I want you to know I understand and appreciate it." He threw down a Signal check, made out to me, for $100,000. I could only reply: "Sam, you don't have to do this." His answer: "I know that, but I want to." I took the check. Sam was one in a million. I thought of Paul Blazer and his "severance" allowance when I left Ashland after, I believe, an equally creditable performance. The contrast revealed the character of the two men.

CHAPTER SEVEN

Creating Momentum at Union Texas Petroleum 1961–67

It was now 1961, nine years from the day we packed our belongings and left Ashland to join Signal and seventeen years after leaving San Francisco to join Ashland. Once again I had missed that elusive ten-year service pin. My California friends asked why I would move to Houston, Texas, when, as they put it, "the traffic was all the other way." I could only reply by claiming to be too young to pick a job based on climate. Since then I have wondered how old I need to get before that becomes a decisive factor. I suspect I could be happy in a tropical jungle, much as I dislike hot humid weather, if I were engaged in an interesting project. Still, I'm sure I'd be miserable in St. Moritz merely making money, no matter how much.

On leaving Signal after the usual farewell dinner, one of my associates borrowed a basketball term to describe me as a "playmaker," which was exactly what the Wall Street owners of Union Oil & Gas Corporation of Louisiana wanted when they hired their new president.

Company Background

The genesis of Union of Louisiana goes back to 1891, when Herman Frasch perfected a method of extracting molten sulfur from its solid ground state, thus beginning the American sulfur industry. In 1896, Frasch formed Union Sulfur Company to market the sulfur extracted from the cap rock of the Sulphur Mines Dome near Lake Charles, Louisiana. From this time until the mid-1920s, Union Sulfur produced ten million tons of sulfur, oil, and gas from this dome. As the world's leading sulfur extraction company, Union distributed its product in France, Germany, and Spain, as well as the United States.

When sulfur production ceased from the Sulphur Mines Dome in 1924, Union turned to oil exploration and production. Its first oil well, drilled on the flank of the dome, was completed in late 1926. With its new emphasis on oil prospecting, the company changed its name to Union Sulphur & Oil Corporation.

In October, 1937, Union's first gas well came in at the Lake Arthur field in Jeff Davis Parish, Louisiana. This also put the company into the business of extracting and marketing natural gas liquids (NGLs). But gas—with a limited market—was a poor cousin to oil. Fortunately this changed by January, 1953, when the W. E. Walker Well No. 1 struck a large gas zone in the Main Camerina sand at 12,300 feet in southern Louisiana.

This discovery, in industry parlance, was a "company builder." The Lake Arthur South field had gas reserves in excess of two trillion feet, with some twenty barrels of light, high-quality condensate per million cubic feet. Condensate is nothing but high-gravity crude oil that drops out as colorless liquid at the separator near the wellhead as the gas is produced. When run through a conventional refinery, it yields mostly gasoline and intermediate distillates, such as kerosene and heating oil. It contains little, if any, heavy fuel oil.

The Lake Arthur discovery was more of a triumph for petroleum engineers than for geologists. Shallow drilling had defined a substantial structure, while deeper drilling had barely touched the lower, high-pressure zones considered too dangerous or too difficult to justify further drilling. After the leases once owned by a large company expired, they were leased by Union of Louisiana. Frank Markel and his company engineers developed techniques for drilling deeper and controlling bottom hole pressures in excess of fifteen thousand pounds per square inch.

Lake Arthur was still under development when I joined the company. One new well at Lake Arthur was almost as good as a whole new gas field elsewhere. Hundreds of billions of feet of gas were packed

The Lake Arthur South field in Louisiana, Union of Louisiana's "company builder." *Courtesy Union Texas Petroleum*

into a relatively small space (only a few acres), within a hundred or so feet of porous sandstone. Pressures like those at Lake Arthur are not "normal," since underground pressures in oil and gas fields are usually roughly equivalent to the weight of a water column (except when the subsidence from the weight of overlying rocks, together with other factors, creates abnormal pressures). "Abnormal" we had at Lake Arthur, for which the company and its stockholders should be forever grateful. Once when I visited the field, where we had pressures of fifteen thousand pounds at the bottom of the hole and twelve thousand at the surface, one of our engineers remarked, "Mr. Marshall, we better hook this thing up carefully or we will have a new earth satellite."

We learned how to handle and use these pressures—but we did not *plan* it that way. Neither the Department of Energy nor the Harvard Business School could have programmed their computers to predict in advance what we learned at Lake Arthur by trial and, luckily, not much error.

Two Mergers

Even with Lake Arthur, Union of Louisiana was an undersized independent, which needed a stronger operating base to "play" as an integrated company against the tough competition. This realization resulted in the almost concurrent acquisitions of two old-line companies in 1960, the year before I joined the company.

Anderson-Pritchard Oil Company was purchased for cash, though the money was not Union's. In the best tradition of that era, the deal was made almost "with mirrors." Eighty million dollars was borrowed from the Chase Bank without recourse against Union and secured only by future production from the properties of Anderson-Pritchard. The small "equity" of some seven million dollars, the amount over and above the monies loaned by Chase, was raised (again in cash) from another oil company through an exchange of certain producing properties in Louisiana. And finally, Union transferred Anderson-Pritchard's pipeline, marketing, and refining operations to newly formed APCO and distributed its stock as a dividend to our stockholders.

In 1961, as Union's newly appointed CEO, I argued against that reorganization. From my refinery background at Ashland, I saw a future for the two relatively modern mid-western plants. But in those days, Wall Street, like many oil companies before and since, held to the view that a dollar was good only if it came from a hole in the ground. Later events confirmed that we had given away what we should have kept. APCO proved profitable, and its refining and other assets were finally sold for more than a hundred million dollars. But at least we "gave" them to our own stockholders and acquired, for relatively nothing, a spread of producing properties and good wildcat acreage located throughout the Midwest and Canada.

Once more, in a minor way, I met myself coming back. I had made a deal for American Independent Oil Company (Aminoil) with Anderson-Pritchard, under which Aminoil carried them for their share of the costs in the independent side of the national Iranian oil company until payout, had paid out. My new company now owned half the net profit in the deal which I had negotiated some years ago for a different company. Even in the oil business, "it's a small world."

The other acquisition, Texas Natural Gasoline Corporation of Tulsa, Oklahoma, followed a more-or-less routine pattern. Voting stock was exchanged for voting stock, which was tax-exempt to the sellers and cash-free to the buyers. The two companies fit like a glove. Texas Natural, a twelve-year-old company built by John Oxley of Warren Petroleum, was one of the first to make a fortune out of processing raw natural gas to yield natural gas liquids. Union of Louisiana already had a substantial interest in two large natural gasoline plants based on its own nearby gas reserves at Lake Arthur; Texas Natural owned and operated seven smaller natural gasoline and LPG plants in Texas and Oklahoma, which produced a million gallons of such product daily. The two together promised scale economies in gas liquification and put us in a position to enter the petrochemical field.

The extraction of liquids from raw natural gas began shortly after World War I. Extraction allowed owners of natural gasoline plants to control substantial quantities of light hydrocarbons without taking the risks of finding them. Generally speaking, only a very few underground accumulations of oil or gas consist exclusively of one or the other. Almost all oil fields produce gas with the oil (casinghead or associated gas), and almost all gas fields produce some liquids with the gas. In addition, the proportion of liquids to gas depends upon pressure and temperature and, thus, is variable. By increasing the pressure or lowering the temperature of raw gas, light hydrocarbon liquids, including ethane, propane, butane, and natural gasoline, can be recovered. All the natural gasoline (and much of the butane) is regularly blended into motor fuel to provide the volatility necessary for easy starting and quick acceleration. Refineries blend as much butane into gasoline as possible—more in the winter and less in the summer—since it is usually the cheapest component of ordinary motor fuel. The light hydrocarbons dissolve into the heavier hydrocarbons like sugar into water. In effect, blending lowers the boiling point in the heavier material and raises it in the lighter material.

In the early days of the oil industry, most of the light hydrocarbons present in raw natural gas were either left in the gas or flared. This made it easy for a natural gasoline plant owner to persuade producers to "dedicate" all their future production of raw gas in return for a share of the proceeds from the liquids extracted and sold. Sam Mosher played this card brilliantly to build Signal in the 1920s. Something, even a small percentage, was better than nothing for a product otherwise wasted. Incidentally, in the early days of the industry, the plant operator took 80 percent of the liquids extracted. Competition being what it is in all phases of the oil industry, the producer's share rose to 40 percent, then to 60 percent, and today can run as high as 75 percent or better. Regardless of the percentage, however, the plant operator ended up owning the right to process the producer's gas for the life of that production. This right was almost like owning a percentage of the producer's reserves in the ground. Sophisticated banks and financial institutions recognized this and willingly provided financing for plants that had enough dedicated reserves behind them.

Consequently, when Union of Louisiana acquired Texas Natural, it acquired processing rights on millions of barrels of light hydrocarbons owned by others. It added millions of barrels of "reserves" to our own already substantial reserves, both in the ground and from processing contracts with other producers. It would be hard to find a better example of how the pieces of the jigsaw puzzle—which is the oil and gas company—can be made to fit together, if you can see the pattern.

New Management for Union of Louisiana

Because of the two mergers, my new company was now one of the strongest independent oil companies in the United States. The company was operated by two stalwart men—Mark Millard, a senior partner of Loeb Rhoades who recruited me, and R. McLean "Mac" Stewart, who represented the Burden interests and was the founder of Austral Oil Company (then a private independent oil and gas exploration organization seeking tax shelters for various wealthy and savvy eastern financial interests, including the Burden family).

Mac, even in his seventies, was an operator in the best sense of the word. He was always on top of every detail of his business. He spent a week or more out of every month in Houston or in the field. His "boys," as he called them, said they had three lives—before Mac came, while Mac was there, and after Mac left. Mark, on the other hand, was a skilled Wall Street "wheeler dealer," also in the best sense of the word. He possessed a creative, imaginative trading mind—if you can't make a deal one way, you can make it by another route. Both Mac and Mark knew full well that the bottom line always measured their success or failure with those for whom they spoke. Since I have always regarded myself as something of an operator and a promoter, working with these two prima donnas was fascinating, to say the least. All of us sat on the board of Union of Louisiana, along with Bill Burden. The other board members were John Oxley, president of our Texas Natural division, Frank Engle, Roger Randolph, Charles Harding, John Loeb, Thomas McCane, H. Harper McKee, and Frank Russell. Bill served as chairman, a rather nice, stuff-shirted spot, which I came to enjoy with other companies I had an interest in after my so-called retirement from direct operating responsibilities later in life. Amusingly enough, Bill was paid a salary—actually an honorarium—which Mark said Bill needed because Mac always kept Bill cash-short while drilling oil and gas wells.

My predecessor as president at Union of Louisiana was Richard T. "Dick" Lyons, a geologist educated at the Massachusetts Institute of Technology. Before joining Union in 1953, Dick had been an executive with Tidewater Associated, an integrated company dominated by J. Paul Getty. I suspect he was wooed away from Tidewater by the same people who lured me from Signal. In my case, I hope they thought they'd hired an oilman—a "playmaker." In his case, I guess they thought they had an oil "finder" with Tidewater's geologic maps in his pocket (or, at least, in his mind). Perhaps once he did have them. But he certainly never found Lake Arthur, as the wife of Frank Markel, the vice-president of production who did "find" it, always reminded me. She was right, and I could only respond by saying I'd been praised for some of my worst performances and criticized for some of my best.

Over the years many of us have witnessed a variety of instances of "paralysis by analysis." Union of Louisiana seemed to have evolved such a system. Early in my tenure as its chief executive officer, I found myself almost overwhelmed with memoranda from our vice-presidents. On inquiring why such extensive paperwork was needed, I learned that administrative procedures required a detailed memorandum in advance before even a vice-president could meet with the president. That rule was changed, and all our top executives were told that only a phone call to my office was required. Routine matters, such as our exploration and production activities, were discussed at regular staff meetings. (Given the nature of this particular art, we called this meeting "the guessing game.")

One of our first moves was to change the name of the company. Since we had recently acquired Texas Natural Gasoline Company and were headquartered in Houston (and since shorter names are preferable to longer ones), we decided upon Union Texas Petroleum.

Renegotiating a Natural Gas Contract

The first challenge I addressed in my new position was to reform a bad natural gas contract. I had to find a way not only to eliminate a thirty-million-dollar contingent liability on our balance sheet, but also to stop that contingent liability from increasing by several more million dollars a year. At stake was a strong financial base for Union of Louisiana (later Union Texas Petroleum).

Long after the passage of the Natural Gas Act of 1938, the Supreme Court unexpectedly reversed a line of previous precedents, which had denied the Federal Power Commission (FPC) the authority to regulate the price of natural gas at the wellhead. Clearly Congress had intended no such thing, as my old colleague from Yale, Justice William O. Douglas, pointed out in a dissenting opinion in the infamous 1954 *Phillips* decision. Nonetheless, the majority opinion becomes law. Still worse for us, such "new" law could be applied retroactively, since the legal theory, once the Supreme Court had spoken, was the "correct" interpretation.

Some years before the Supreme Court came forth with this interpretation of the Natural Gas Act, Union of Louisiana had entered into a contract to sell all the gas it produced in most Louisiana parishes to Texas Gas Transmission Company (an interstate gas pipeline company) for twenty years, at a starting price of nine cents per thousand cubic feet. Under the contract, the starting price was to escalate, as the prices paid by others under later contracts might increase. The FPC attempted to invalidate administratively all such "most favored nation" escalator clauses, but this ploy was challenged in the courts. During this period, Texas Gas Transmission had paid for gas in accordance with the terms of its purchase contract, which was subject to Union's liability to refund the difference between the starting price of nine cents and the current escalated price of about nineteen cents per thousand cubic feet, if the FPC prevailed. Good old Mac Stewart bellowed that they could not do this to us—it was unfair, illegal, unconstitutional, and impossible.

I finally convinced him that we could not afford to risk losing if we could settle on a decent basis. Mac, somewhat reluctantly, agreed to let me try, and a personal approach to Bill Elmer, head of Texas Gas Transmission, indicated a willingness to settle. He too preferred to avoid the risk of losing—though, in his case, not so much regarding the issue of price (utilities can generally pass on price increases to their customers) as on the question of assuring a firm, long-term contract of some two trillion feet of gas from the big Lake Arthur field in Louisiana.

After months of bargaining, Bill and I agreed on a ten-year contract at a flat price of 16.25 cents per thousand cubic feet for Lake Arthur gas. Much as I disliked selling anything in an inflationary era at a fixed price for future delivery, it seemed a lot better than nine cents, which we might have been forced to take. Moreover, we were to be relieved of all obligations to make refunds for alleged past overcharges, except about three million dollars' worth. The "after tax" effect of even the three million cut that figure in half, and our thirty million dollar contingent liability was gone.

Under the proposed settlement, all calls on future production in Louisiana outside Lake Arthur were released by Texas Gas Transmission. Without this release, we could not afford to develop additional gas in Louisiana if we had to sell it for sixteen cents, let alone for nine cents. Both the FPC and its successor, the Federal Energy Regulatory Commission, regularly refused to recognize the foregoing simple application of elementary economics, which is, quite probably, one of the most important reasons for the severe "shortages" of gas experienced during several winters in the 1970s. No one has yet found a way to balance supply and demand through increasing demand by forcing a commodity to be sold at a low price, while rendering it less and less profitable to increase supply.

The settlement between Bill Elmer and me was only the beginning. More than twenty local distribution companies supplied by Texas Gas had to agree, as did their local regulatory authorities, the staff of the FPC, and the commissioners themselves. After thousands of miles of air travel and months of negotiations, all of these many conflicting interests decided it was best to settle. We almost wore out a set of engines on Union's Lockheed Lodestar, a company plane I found we owned when I joined the company. At least for once, a company plane earned its keep. Ours were no pleasure trips; once, during settlement travels, a bad squall line an hour out of Houston turned us back. We all felt lucky to return safely.

I have often wondered whether we would ever have climbed the last barrier to a settlement had it not been for a bit of good fortune. In the summer of 1933, during the early days of the New Deal, three young, idealistic bureaucrats shared an inexpensive rented house in Chevy Chase, then a suburb far out of Washington, D.C. I was one of them, Norman Meyers was another, and the third was Joseph Swidler, destined to be chairman of the FPC at the time of the settlement negotiations. Joe was a dedicated public servant, and neither friendship nor influence could sway him. But our old association got me an appointment to argue our case on its merits. I remember starting informally by recalling one of our favorite stories from the Chevy Chase summer and then going into the meat of the arguments, which prevailed with Joe and his fellow commissioners. Perhaps even the FPC didn't relish a court test, where it might have had the law on its side but the opposition would have the equity. For all the parties, as with most settlements, a part of a loaf is better than no loaf at all.

Marketing Gas Liquids

Once Anderson-Pritchard and Texas Natural had been acquired, the newly named Union Texas Petroleum was remade into a modern, broad-based oil company. Still, its degree of integration left much to be desired. The two lightest hydrocarbons extractable in our plants, propane and ethane, presented serious marketing problems. It was easy to sell propane (bottled gas) at wholesale in the winter when heating demand peaked, but summer marketing was difficult. It was so difficult, in fact, that sometimes propane was given away at the plant merely to earn tank-car mileage to move it to someone who could use or store it. In many other cases, propane was flared at plants, which had to process gas on a year-round basis regardless of demand for some of the products. In a very real sense, this situation was another illustration of the economic forces generated by the law of capture. If any one oil producer, drawing from a common reservoir, elected to produce, all others must do likewise, regardless of what happened to any of the light hydrocarbons produced with the oil. Unless all awaited the advent of winter, none could afford to suffer physical drainage at the hands of those who either could not or would not curtail their production.

Economic pressures arising out of the law of capture forced Union Texas, as it had our competitors, into local and retail propane marketing through its subsidiary, Texgas. Devoting some of our always limited capital resources to propane marketing had to be done over the objection of our financial experts, who argued that we could not show an adequate return on such investments. We should confine ourselves, they said, to investments that showed at least a 15 percent return after taxes. In theory they were right, but it depended on how you figured the return. They started with what they assumed to be a normal margin between bulk prices at the plant and wholesale and retail prices to consumers. They conveniently overlooked what needed to be compared—not an assumed normal margin but rather, no profit for part of the year. The fact was that, unless you had made investments to develop some captive market, for half the year there was no margin at all between plant prices and retail prices—the stuff would be flared and you got nothing.

With Signal, I had been down this road before. Much as we all might have preferred to sell at the plant gate and be done with it, there was, for us at least, no viable alternative but to acquire captive markets near the consumer's burner tip, and this we did. Some years later, after Union Texas merged into Allied Chemical, their equally financially minded experts almost persuaded Allied's board to sell its propane marketing operation. At the last moment, somebody realized that the cost of not having a market exceeded the gain from the sale of the physical assets—this was yet another illustration of the difference between a "figure" man and an "oil" man. Even in the best large organizations, it is a tough task to persuade people to think in "integrated" terms.

In a still larger arena, both the Federal Trade Commission and the Antitrust Division of the Department of Justice have almost universally regarded "captive" markets as inherently wicked. Perhaps they are, if designed solely to control competition. But what, for example, is an oil refiner to say? What is a gasoline marketer or a producer of light hydrocarbons supposed to do? Because of the law of capture, they must produce their supplies of oil and raw gas every day. With oil and its refined products, they can store some but not for long. Light hydrocarbons can be stockpiled, if at all, only in relatively small amounts and only for a very few months. There are economic pressures, different from those in most other industries, that seem to pass unnoticed by governmental regulators bent on proving that oilmen and oil companies are a bad lot.

So much for the propane marketing problems of Union Texas. What about ethane, the next lightest hydrocarbon? It is even more difficult and expensive to liquify and hold in a liquid state, which is both good and bad. It is good because you can leave it, unlike propane, in the gas stream without risking condensation to retard the flow of gas through a pipeline. It is bad because unless you extract it as ethane, you are selling it on either a volumetric or heating-value basis rather than for somewhat higher values as a petrochemical building block. Both of our big plants in Louisiana had relatively high concentrations (5 percent or more) of ethane in their gas streams, most of which was left unextracted and sold either by the cubic foot or on a BTU basis to be burned as fuel. We needed further downstream integration, another captive market, this time in the petrochemical business, to upgrade the value of our ethane. We set about acquiring such a market, which eventually led to a merger with Allied Chemical.

International Exploration

When I joined Union of Louisiana, we produced fifteen thousand barrels of crude oil and two hundred forty-five million cubic feet of gas per day. Domestically, Union's production was concentrated in southern Louisiana, with some small interests in Texas. Back in 1956, the company had entered into the international arena as part of the TULM consortium, made up of Union, Tennessee Gas Transmission, Lion Oil Company, and Murphy Petroleum. Their first concession in Lake Maracaibo's Marlago field in Venezuela resulted in significant oil production. Another venture, in Bolivia, found natural gas but no market to justify commercial production. Union also had a Canadian office in Calgary, Alberta, which managed moderate production.

I knew the importance of the international market. It was no secret that the United States was the most mature oil province in the world. I knew Lake Maracaibo well from my Signal days. I also had experience with Signal in Argentina, where Mark Millard, Henry Holland, and I had negotiated a production-sharing contract with the state oil company that paid off handsomely with the Mendoza oil field, an "elephant" by yesterday's or today's standards. Although not as successful as the other two, my first taste of international oil exploration and production had been in the Kuwait neutral zone while representing Ashland.

In 1965, Union was approached by Asamera, Ltd., a Canadian company headed by Tom Brook, to take a 30 percent interest in their Indonesia concession. Our exploration vice-president, Elliott Powers, strongly recommended that we join. While many other oil companies would have taken weeks (if not longer) to decide, in a matter of days (after being satisfied about the geologic prospects and political stability of the country) we committed $1.5 million to enter the deal. The ability to move quickly was a major competitive advantage.

The payoff in Indonesia would come several years after I left the company. In 1968, Huffington, Inc., and Virginia International entered into production-sharing contracts covering 4.3 million acres in Sumatra and East Kalimantan. They went looking for partners, and Union Texas took a 35 percent interest. In the early 1970s, significant oil and gas deposits were located in East Kalimantan. Twenty years later, this property is still a prize for Union Texas.

The British North Sea was another venture that had a big payoff for Union Texas. In 1965, we received an exploration license in the North Sea, which we farmed out to Shell and Exxon after the conservative—too conservative, as I'll argue—management of Union Texas/Allied Chemical decided they did not want to drill. In return for the wildcat well, each got a 25 percent interest to go along with our 50 percent interest. The result was the discovery of the large Sean gas field in 1967, the year I left. Of equal or greater value was our experience in the region, which led to two bigger finds in the next decade for Union Texas: the Piper oil field in 1973, and the Claymore oil field in 1974.

Elliott Powers, our vice-president of exploration and production from 1956 to 1970, credits my tenure at Union for creating the momentum that would lead to these later successes for the company. In any case, I was certainly more of a risk-taker than my predecessor. Perhaps that's why Mark Millard and Mac Stewart went to the trouble to hire me. My philosophy was to keep making a lot of deals, and I had confidence that we would make more good ones than bad ones. I think we did.

Merging—and Leaving—Union Texas

About two years into my tenure at Union Texas, my top priority became developing markets for the light hydrocarbons, which the company owned and controlled in substantial volumes. I initiated conversations with Kirby Fisk, then chairman of Allied Chemical, to enter into a joint venture of a petrochemical facility in Geismar, Louisiana, that would be owned in equal shares by Allied Chemical and the stockholders of Union Texas. We ran into the transfer-price problem of how to value raw materials, which Union Texas would supply to Allied to make finished products, such as toluene, benzine, and other light hydrocarbons. This got so complicated that Kirby and I came up with the idea that it would be simpler to merge the companies and stop arguing about transfer prices. We made a "handshake" deal, since we were the only two persons who knew about it. If there was a leak, it had to be one of us. News of the deal never leaked until the very end. When I called John Oxley, a major stockholder of Union Texas, to tell him about it, before I could get started, he said that a reporter was waiting in the office who claimed that a deal to merge Union Texas and Allied Chemical had been made. John had already told the visitor that I denied it, so I told John to "undeny" it because it was true. John blew up but got over it quickly.

Kirby finally agreed to the price I wanted for Union Texas, on the condition that I remain as president of their new subsidiary, Union Texas Petroleum Division, for at least five years. In this way my efforts would avoid the dilution of the value of Allied stock. As a large stockholder of Allied myself, I agreed, and the merger was consummated in February, 1962.

I was ready to leave after the five-year term (1961–65), but we were involved in various projects that promised to make substantial amounts of money for both the company and me. I ended up staying for two years longer. By this time Kirby had died and had been succeeded by John T. "Jack" Conner. Without Kirby, the synergy between Union and Allied was never the same for me. In 1967 Conner offered me a retirement pension from Allied, if I would sign an agreement not to involve myself in any venture where a conflict of interest with Allied would arise. I readily agreed, since this was good business practice anyway. To this day we have never had an occasion to evoke the contract, and I still draw an Allied pension, which I think I earned.

One story illustrates why I left. After negotiating for two years to build a petrochemical facility at the head of the Persian Gulf, some of my enemies at Allied ran around my flank to explain to Jack that it was a bad deal. I defended the project by explaining that if we can't make money paying thirty-six cents a million BTU for gas going into an ammonia plant, we ought to get out of the fertilizer business. His reply was, "What is a BTU?" I remember thinking: here is the chief executive of a major chemical firm who does not know what a British Thermal Unit is. I decided then and there I would not enjoy working for this man and that the quicker I could get out the better.

It was time to go into the oil business for my own account. My six-year association with Union Texas would be my last employment with a public company. I was now on my own.

CHAPTER EIGHT

On My Own, 1968–Present

With my sixty-third year on the horizon, I was not ready to go to pasture, though I had a full career behind me. I had been in government service for an exhilarating five years (1933–1935, 1941–44) and in corporate life for over a quarter century with Standard of California (1935–37), Pillsbury, Madison, & Sutro (1938–41), Ashland Oil & Refining (1944–52), Signal Oil & Gas (1952–60), and Union Texas Petroleum/Allied Chemical (1961–67). My financial house was in order, but I was still a workaholic. My bad leg diminished my tennis and golf recreation, but my mental energy seemed as plentiful as ever. There were a lot of adventures in the oil patch at which I needed to try my hand. Not all would be successful, but the next quarter century would have its rewards.

Independent Refining Corporation

None of us is ever wholly successful, and certainly I have not been. One of my greatest failures sprang from my misjudgment of a key employee. For years there has been a superstition in the oil business that the only way to make money is in production, such as a lucky wildcat strike worth hundreds of millions of dollars. In my spotted career, I have never found that completely true, for I've seen great fortunes made in independent refining and marketing. Such is the exception to the rule, but it does happen. It was certainly true in the case of Ashland, as chronicled by Otto Scott in *The Exception*, a history of Ashland and Paul Blazer. The book describes how Ashland almost became a major by being very successful in refining and marketing. I. A. O'Shaughnessy made a fortune with Globe Refining Company in Chicago. Old Colonel Barton of the Lyon Oil Company, later acquired by Monsanto, was a successful downstream entrepreneur. There have been several West Coast refineries that proved successful in these areas, such as the Wilshire Oil Company, headed by George Machris. His trick was to keep his feedstock costs down by violating all the proration agreements in California.

After I left Union Texas/Allied Chemical, I had the notion to put together an independent refining and marketing group. At the time the federal government was subsidizing small domestic refiners by granting them more entitlements to process lower-priced oil than larger refineries. This program sprang from the circumstance that, with some crude-oil categories price-regulated, and other categories not regulated, refineries had wildly disparate feedstock costs. To equalize these costs, refineries with crude costs lower than the national average were required to buy the right to refine their crude from refiners with higher-than-average costs. Furthermore, small refiners could preferentially sell more entitlements or buy fewer entitlements in order to run low-priced crude. The "small refiner bias" under the Old Oil Entitlements Program, which began in November, 1974, continued the small-refiner subsidy that began in 1959 under the Mandatory Oil Import Program.

In the original regulations, the subsidy per barrel for a ten thousand barrel-a-day plant was $0.62. This increased to $1.79 per barrel by May, 1976, and would fluctuate between $1 and $3 per barrel in the following years. Some of us thought that, rather than try to fight the regulations as our free-market instincts might dictate, we should take advantage of the opportunity.

In 1975, I founded Independent Refining Corporation to acquire a small refinery located in Winnie, Texas. It was rated right at ten thousand barrels a day, which qualified for the most entitlements.

I knew the refinery well. I had bought it for Allied when its previous owner, Glenn McCarthy, the famous Texas wildcatter, went broke. Along with Allied, it was owned by Standard of Indiana, and some investment bankers in New York led by Loeb Rhoades. I had to purchase the refinery under the "cover of darkness" because I was sure that Allied would not sell it or would, at least, raise the price, if they knew I was sponsoring the purchasing group. Thus, through the closing I was only known to the banks and the selling group as "Mr. X." Allied found out I was involved only after I had purchased the refinery for a pittance, the price reflecting the disrepair of the purchased assets.

The major inducement to the project was the small refiner subsidy, but another attraction was some nearby crude oil production, which yielded particularly good aromatics when refined. The posted price for this crude was less than it should have been because its owner, Monsanto Chemical Company, was paying a transfer price to supply its nearby Chocolate Bayou refinery, which extracted aromatics and various petrochemicals. I wanted the same crude to extract my own aromatics.

After I acquired the refinery and moved to buy the crude, Monsanto ceased taking bids and kept its crude for Chocolate Bayou. This, I guess, should not have surprised me. I then had to figure out how to supply the refinery—and run it. Allied had more or less neglected it because, in their league, it was too small.

We began by improving the refinery. The first steps were to enlarge our storage capacity at the plant and improve our pipeline connections between Winnie and the Deepwater facility on the Gulf of Mexico, where we brought in crude from tankers. We obtained some cheap foreign cargoes and made enough money off our early runs to purchase a used catalytic cracker.

We had a profitable operation for the first four or five years. Things quickly soured when it became apparent that lax financial controls and a dishonest president had sapped the financial strength of the enterprise. A management reorganization during the second half of 1980 proved too little too late when the entitlements subsidy evaporated in early 1981. In September, 1982, I sold the refinery to a foreign company at a loss and took the tax deduction. The refinery never turned the corner, and it was liquidated the next year.

My experience with Independent Refining was a sad chapter because the refinery had had such potential. Had we integrated the cat cracker, the refinery could have been profitably run for another decade, at fifteen to twenty thousand barrels a day. While that's not world-class, the refinery was well-located on the Gulf and tied to most of the major midcontinent pipelines to move the products to market. I'm sorry I picked the wrong person to run it. The guy was smart enough, but there's no substitute for integrity—you either have it or you don't. He didn't, and that's the nicest way I can put it.

Back to Exploration and Production

My first love in the oil business has always been exploration and production, but, in terms of making money, my bread has been buttered more in refining and marketing. Soon after I left Union Texas/Allied Chemical, William Moss, an independent producer from Texas, and I organized a corporation named The Petroleum Corporation (Petco), which, as the general partner, organized and managed a series of private placement, limited partnerships to search for oil and gas. The defining characteristic of our partnerships was that Petco would treat the limited partners like the oil companies treated each other. As general partner, Petco would take some of the risk and run the drilling programs professionally rather than as a promoter. We would only invite to join us those people we knew and who knew and had confidence in us. Beginning in 1970 we ran a succession of private syndicates (organized as limited partnerships), with William Moss Properties, Inc. and me as the general partners.

The general partners let the limited partners take the original risk of financing the assemblage of acreage blocks and drilling the discovery well. When development started from the commercial finds, the limited partners and the general partners shared the cost. Bill and I had a quarter interest together, and the limited partners had a combined three-quarters interest. The way the tax laws were written at that time, almost all the money we put up qualified as an intangible expense deductible from income taxes, except the original risk, which was not our money in any case.

Bill Moss was a superb trader who had connections throughout the oil industry. I was more attuned to the details of our programs. I suppose our different talents and interests made our ventures successful.

Over the years the program got bigger and bigger. I suspect that, between 1970 and 1986, Bill and I spent several hundred million dollars drilling wells in West Texas, Oklahoma, and Louisiana. In 1984 we placed our properties into a new corporation, The Petroleum Development Corporation. When the federal tax statutes changed to take away traditional incentives for exploration and production, culminating in the Tax Reform Act of 1986, Bill and I sold Petroleum Development to Presidio Oil Company. In March, 1987, we took publicly traded Presidio stock for our oil and gas interests, and two months later I became a director of Presidio, which I remain today.

Presidio Exploration Company was incorporated in 1968 and went public as the Presidio Oil Company nine years later. Our sale was one of six acquisitions that Presidio made between 1987 and 1989, for a total of $420 million. They felt they were buying at the right time and believed they could operate the acquired properties more efficiently by buying clusters of properties in the same general area. George Giard, Presidio's chairman, may be the only Rhodes scholar to ever grace the oil industry. He's a good oil executive and sound financial man.

In addition to my Presidio interests I held with the limited partners, I had a variety of production interests of my own. Much of this resulted from the Signal and Union Texas/Allied arrangements, where I could do any oil deals I pleased as an individual as long as they did not conflict in any way with my basic responsibilities to the corporation. In 1984, I consolidated my personal interests, along with some of the production interests that Bill Moss and I shared outside of Presidio, into a company which I named after myself. Marshall Petroleum, Inc. is strictly a family corporation. We are not as active today as in the last decade, but under the direction of our vice-president, Henry Schlesinger, we occasionally participate in drilling ventures. Marshall Petroleum is where I've hung my hat for almost a decade, and it's also where I'll work my last day.

One of my important activities of the last decade has been serving as a director of two very successful corporations—Koch Industries and The Coastal Corporation. Both associations have been personally and professionally rewarding, and, in the case of Koch, I've found some high adventure as well.

Koch Industries, Inc.

Koch Industries is an independent oil refiner, marketer, and transporter located in Wichita, Kansas. In terms of sales, it's the largest privately held energy firm in the country and the second largest

The present staff of Marshall Petroleum, Inc. *From left, standing:* Patty Oxley, Donna Roebuck, Eyvonne Scurlock, Henry Schlesinger; *seated:* Dan Manning and Howard Marshall.

private company period. It's my good fortune to be a part owner and director of this well-run, profitable company.

The genesis of my involvement with Koch goes back to my friendship with Fred Koch and my interest in the Great Northern Oil Company, which I cofounded in 1954. I got to know Fred during World War II, when I was chief counsel and assistant deputy administrator of the PAW. He was a brilliant petroleum engineer with entrepreneurial instincts, a rare combination that would make him very successful.

In 1925, Fred cofounded Winkler-Koch Engineering with a Wichita engineer, L. E. Winkler; the firm would eventually patent the thermal catalytic cracking process. In addition to setting up refineries around the world, the company protected their patents well and received substantial settlements from violators. In 1940, Fred cofounded and ran a midcontinent oil company called Wood River Oil and Refining Company. I. A. O'Shaughnessy and Hank Ingram were also Fred's partners in the firm, whose major asset was a refinery across the river from St. Louis, Missouri, which produced 10,000 barrels per day. In 1946 Fred bought Rock Island Oil and Refining Company, which added a refinery and oil pipeline gathering network in southern Oklahoma to his holdings.

The story of how Fred Koch purchased an interest in Great Northern Oil Company illustrates his business stature. In 1959 a refining executive at Sinclair, Ike Moore, alerted Fred that Sinclair's 35 percent interest in Great Northern was for sale. (Sinclair had purchased its share from Southern Production Company some years before.) Fred approached Percy Spencer, the head of Sinclair (whom he knew personally), about it. Percy stated that his company would take book value, and Fred, who always said he jumped at the chance to buy a refinery for book, didn't hesitate. He told Percy he would buy it, although he didn't know anything about the facility. Actually, Fred had investigated it very thoroughly; maybe he didn't know as much about it as some of us, but he knew enough. He had a sense of strategy and recognized that the refinery was in the right place at the right time. It had access to a relatively good supply of crude and was tied into a major market for products. The Twin Cities area was the ideal place. There were no major refineries there except for a small, inefficient one run and owned by the Erickson family of Minneapolis.

When Percy asked Fred how he wanted to pay for it, Fred asked if his check would be all right. The check was for about five million dollars. Percy could only shake his head and mutter that he didn't know there were any fellows like that left anymore. I later asked about it: "Fred, don't tell me you have five million dollars lying around in a bank somewhere." He responded: "Of course not. I just called up First Chicago and said I just spent five million. Send me a note." That was the kind of company and reputation Fred Koch had back in the 1950s.

Fred swung the deal by using his great competitive advantage—paying cash. Later, when Fred's son, Charles, bought Sun's Corpus Christi Suntide refinery, he again won by a large cash bid—$265 million to be exact. Sun needed the money and thought they were selling a marginal operation. The refinery was running about fifty thousand barrels a day when Koch bought it. After new facilities were added and other operational changes made—changes that the refinery's previous owner hadn't thought of—within eighteen months the refinery was running well over a hundred thousand barrels a day. I remarked to the Koch people that at Great Northern, we could run the lousiest crude in the world and make high-quality products. With Suntide, we ran the best crude in the world to make the best products commanding the highest price. It has been run successfully since it was purchased in November, 1981, and all we are doing now is expanding to 250,000 barrels a day. But we're getting ahead of the story.

As explained in chapter 6, I held 12 percent of Great Northern, which was less than my original 20 percent because of some options exercised by A. G. Becker, an investment baking firm in Chicago that financed my purchase. Fred Koch and I would buy any shares of Great Northern that Becker placed on the market, stock that originally belonged to me, and, over the years, I built my 12 percent back up to about 16 percent.

In addition to my 16 percent and Fred Koch's 35 percent, the balance was owned by Pure Oil Company, which previously had bought out Woodley Petroleum. We pursued Pure's interest and negotiated with their president, Bill Milligan. When Fred and I previously decided to expand Great Northern, we had met with Pure and Milligan. They had about 49 percent of the stock versus our bare majority. Fred made a logical case for aggressively expanding capacity, and Milligan responded by saying that the industry needed to cut refinery runs instead. I said to Bill: "I thoroughly agree, but it is the other guys' runs I want to cut." Pure finally realized that our argument was not to increase capacity for its own sake but to profitably take care of our customers' future needs.

When Milligan shared our argument with Pure's board, J. R. Parten, a director, chimed in that he had no doubt Koch and Marshall would run the enlarged refinery at capacity. It was one of Fred's partners, Hank Ingram, who once said the way an independent refinery makes money is to run it as hard as it will go. That is about what we did with Great Northern, despite Pure's conservative inclinations.

In 1959, Wood River was renamed Rock Island Oil and Refining Company. Charles Koch, who joined Rock Island in 1961, became president in 1966 and assumed the chairmanship upon the death of his father a year later. One of his early acts was to rename the company Koch Industries, Inc., in honor of his father. Charles had all of his father's ability plus some. He earned his Master's degree from MIT in both chemical and mechanical engineering. He had worked at Great Northern in his mid-20s and knew the refinery better than the refinery manager did.

Several years after Fred's death in 1967, we had to figure out what to do with my stock interest. Union Oil of California (now Unocal), which had acquired Pure in 1963, wanted to sell their interest. They even suggested to potential buyers that they could persuade me to join them to get control from Koch. I knew better than to do that, so Charles and I decided to trade my Great Northern interest for Koch stock. Charles paid me a great compliment after the stock swap when he said: "I generally do not like partners, but Howard Marshall is an exception." I think it is one of the only times he ever consummated such a trade with an "outsider." It was either the smartest or the luckiest thing I ever did, and maybe a combination of both. It turned out to be the best deal I ever made.

Charles gained full control of Great Northern for Koch when he bought Union's interest in 1969. Union was experiencing some lean years, and Fred Hartley, their chairman, wanted cash. Koch bought out Union using Union's credit, but Union was able to show a handsome profit for the year because the sale price was well above book value.

Today Koch Industries has grown into one of the largest corporations in the world. Koch has always been courted by investment bankers to go public but has resisted them. One of the reasons Koch could do so was because the company was strong enough to build the business without them.

Koch is a large energy company, but not a "major," since it's not proportionately strong in exploration and production. It is strong and smart in virtually every other phase of the oil business, including all aspects of chemical technology. Koch is also active with natural gas and its derivatives, as well as minerals and agriculture. The company is divided into eight groups: crude oil and oil field services, hydrocarbons, refined products, exploration, carbon, chemicals, chemical technology, and real estate.

One of Koch's advantages is that, with private ownership in the Koch family—and maybe I am an honorary member of it—we don't have to be concerned with issues affecting a publicly-held company. We do not publish figures and bare our soul regularly through filings with the Securities and Exchange Commission (SEC), the New York

A Koch Industries board meeting with Marshall (*far left*) and David and Charles Koch (*at end of table*). *Courtesy Koch Industries, Inc.*

Stock Exchange, and other public bodies. Most of the major oil companies court publicity, but all Koch worries about is serving its customers and reinvesting the proceeds, which is the way the capitalistic system is supposed to work.

Everyone on the Koch board knows the business or, at least, important parts of it. We can hold a board meeting over the telephone and do so regularly. We debate things until serious differences are resolved. That has been true for most of the company's history, except in the early 1980s, when a family feud erupted that almost took the company public. The dispute centered around one of the brothers, William Koch, who demanded an unearned role in the company and wanted to take money out of the corporation in a manner not in the best interests of either the company or its other shareholders.

This struggle for corporate control reminded me of a similar tiff many decades before. When I was executive vice-president of Signal Oil and Gas Company on the West Coast, the company was controlled lock, stock, and barrel by the Mosher family. Sam Mosher, a great entrepreneur, held each annual meeting in his office. I still remember seeing him lean back in his chair and state: "We do not need to wait until the tellers count the votes. I can assure you a quorum is present." All he meant was that he was there.

But in 1924, Mosher's partner, Robert Bering, secretly set out to gain a majority of Signal's stock. Mosher found out in the nick of time and set out to do likewise. The race came down to a hundred shares. Mosher located a widow in San Francisco, who owned this balance, and visited her to negotiate. After his offer was accepted, she went to retrieve the certificates. Just before she handed the stock to Mosher, she hesitated with the words: "My husband always told me that, in a situation like this, I should have a certified check." By this time, Mosher saw Bering coming to the door. Mosher, who was quick on the draw, said: "Oh, pardon me madam," and turned the check over and wrote on the back, "personally certified by S. B. Mosher." She took the check, Mosher got the stock, and Bering was turned away. That was the difference between success and failure for Mosher and Signal.

I remembered that lesson when my oldest son, Howard III, to whom I had given half of my voting stock in Koch, threatened to vote his stock interest in favor of the dissident group and against me. He held a small amount of the total stock but a critical percentage of the voting stock. I made a case based on family obligation and loyalty to get him to stay with our group without success. Charles asked, "What do we do now?" I answered, "I'll tell you what I will do tomorrow. I'll offer Howard eight million dollars for his Koch stock. I'll also tell him that I helped build this company and am proud of it. It should remain private to maintain its competitive advantage."

Howard accepted my offer the following morning. William Koch reportedly offered to buy Howard III's interest for a higher price, but I convinced Howard that he did not have any assurances that William's offer was any good. I handed him my check and said, "This check is good and you know it."

We do not have any dissidents anymore. We have meetings that discuss the issues, the problems, and the opportunities. We decide whether to go or not go—and generally we go because we dissect the matter to the point where sensible businessmen could not disagree. Charles is always talking about competitive advantage, and one of our competitive advantages is that we can pay cash for the assets we need. When we borrowed money from a syndicate (led by J. P. Morgan) to buy out the opposition, William Koch, the leader of the dissidents, said that our hands were tied, and we could never make another deal.

Oscar S. Wyatt, Jr. (*center*), founder of Coastal Corporation, enjoying a good laugh with fellow directors Howard Marshall (*left*) and E. O. Buck. *Courtesy Coastal Corporation*

He forgot that we could pay off the note—which we did. Now, instead of having disagreements in the board meetings, I look forward to them. I'm always anxious to know what we have on the cooker.

The Coastal Corporation

Oscar S. Wyatt, Jr., is one of the most interesting people I have ever known in the oil business. Oscar is always thinking ahead and knows what he's doing. He's worked through every phase of the business. I had done business with Oscar when I was president of Union Texas, but I didn't know him well at that time.

Oscar is a hard-driving businessman and a colorful character. He is also controversial to many. An April, 1991, article in *Texas Monthly* on Oscar is entitled, "Meaner Than a Junkyard Dog." To such critics, Oscar has a pat answer: "My stockholders do not pay me to run a popularity contest—they pay me to make money," to which I always reply: "Yes, Oscar, and you are the largest stockholder of the company." Oscar is an interesting, astute individual and one of the best traders I've ever known.

The genesis of my involvement with Coastal (then called Coastal States Gas Corporation) concerned a settlement Oscar and Coastal made with the Texas Railroad Commission and the Securities & Exchange Commission in the early 1970s. Lo-Vaca Gas Gathering Company, a wholly-owned Coastal subsidiary linking South Texas gas producers to San Antonio and other Texas towns, curtailed deliveries to several natural gas distribution companies beginning in 1973, despite contract provisions stipulating firm service. The problem was that Lo-Vaca was sandwiched between fixed selling prices and escalating gas-purchase costs. In a 1977 settlement with the Texas Railroad Commission, Coastal agreed to divest the subsidiary (renamed Valero Corporation), refund $1.6 billion to ratepayers, and spend several hundred million dollars to find and develop gas reserves to sell to Lo-Vaca's customers at discounted prices. The settlement with the SEC also required Oscar to appoint a majority of outside directors to the Coastal board.

I got a phone call one day from Oscar. He explained the SEC decree, inviting me to be one of the outside directors. I politely declined with these words: "Oscar, I have enough to do. I am not collecting directorships." He countered: "Well, would you do it just for me?" I said that, on that basis, I would. He closed: "If the SEC calls you up to test your qualifications, don't admit that you know me very well." I said that was easy, since I didn't.

I received the call from the SEC. They went through my spotted career and asked various questions. I quickly got to the point: "If I become a director, I will do as good a job as I know how and use balanced judgment with no favoritism." The regulators had the court approve my appointment as a director of Coastal and chairman of its executive committee.

My first board meeting was on October 2, 1973. The settlement stipulated that Oscar was bound to the new directors for a year. There were six other outside directors, one being my good friend and personal banker, E. O. Buck, from Texas Commerce Bank. The others were Harold Burrow, Charles Jones, Edward Mosher, Warren Smith, and Fletcher Yarbrough. At the end of the year, Oscar invited us to stay on, and by this time I had found Coastal so interesting that I agreed. Everybody else did too. I think it has been one of the most stimulating boards I've ever sat on. In contrast with Allied Chemical, where the board was largely composed of representatives of the financial community (who were there mostly to generate stock and bond business), the Coastal board was learned and productive.

I have stayed on the Coastal board continuously since 1973. I was chairman of the executive committee until November 7, 1989, when Oscar relinquished his title of chairman. In these years I have watched the company expand, and its growth, though different from Koch Industries, has certainly paralleled that company's.

The oil industry is full of modest beginnings, and Oscar Wyatt and The Coastal Corporation certainly qualify. In 1951, Oscar mortgaged his car to borrow $800 to invest in some gas wells in South Texas. The Hardly Able Oil Company was as apt a name as any for his company. Two years later, Hardly Able was renamed the Wymore Oil Company to run its growing (but still modest) South Texas properties. From this base, it became necessary to expand into gas gathering to find a market for the gas. The bigger pipelines at the time were not interested in Wymore's "scraps."

On November 10, 1955, Coastal States Gas Producing Company was incorporated with assets of sixty-eight miles of small-diameter, natural gas gathering lines and thirty-nine oil and gas wells. By entering into contracts to buy gas at one point and sell it at another point to lock-in margins, financing became available to construct gathering lines. This formula resulted in rapid growth, and, within five years, the number of gathering systems had grown from five to thirty-eight, with capacity exceeding over one billion cubic feet a day. Wyatt had found his niche: aggregating small quantities of gas, which his larger competitors regarded as crumbs on the table.

From this base, Coastal has expanded into a variety of new and complementary energy areas across both the United States and the world. The new areas were liquids extraction (1959), intrastate gas distribution and crude oil gathering, pipelining, and refining (1962), gas liquids transportation (1965), interstate gas transmission, chemicals, and coal mining (1973), oil terminaling and international marketing (1975), fuel oil marketing (1977), commercial trucking and cogeneration (1985), gasoline marketing and convenience stores (1986), and foreign refining (1991). In 1973, the company became Coastal States Gas Corporation, and, on the first day of 1980, its present name, The Coastal Corporation, was adopted.

I can surmise Coastal's growth by looking at the value of my own stockholding in the company. My nominal purchase of five thousand shares for $47,000 in 1973 is now worth three-quarters of a million dollars. This appreciation directly reflects the profitability and growth of the company. In the year I joined the company's board, $38.2 million was earned on sales of $788 million. In 1993, earnings were $116 million on revenues of $10 billion.

It is a monument to Oscar Wyatt that he regrouped from his troubles of the early 1970s to build a very successful company. His success is what the capitalistic system is supposed to generate. You hear the stories that Oscar Wyatt was and is dishonest and devious, but they are highly exaggerated. Once you get Oscar Wyatt's word, you can bank on it. You do not have to run around and force him—he will do what he says. He is tough, tireless, and opportunistic, and he does keep the lawyers busy—but that is more a reflection of what it takes to succeed in today's business environment than on Oscar.

Once a lawyer friend in California, Al Weil, head of Mobil's West Coast subsidiary, was being criticized by a business rival. I was representing Standard of California (now Chevron) at the time and moaned to Weil about it. Al looked at me and said, "Howard, you don't expect your competitors to love you, did you?" That is what I mean about Oscar Wyatt. He has been a highly successful competitor against people that have not been half as successful. They do not love him—they fear him, and they fear him because he is a better businessman than they are. Oscar has always been absolutely straightforward in his personal conduct toward me, never undependable, devious, or dishonest. With Coastal and Oscar Wyatt, I stumbled upon another mutually rewarding relationship.

American Petroleum Institute

A third board I have sat on that has been a window to the integrated petroleum industry has been the American Petroleum Institute (API). API is the successor to an industry-government organization put together in World War I to assure petroleum supplies for the armed forces. When the National Petroleum War Service Committee was reorganized as API in 1919, it became the largest trade association in the oil industry. Its members were both majors and independents, but this changed, in 1929, when many independent producers broke away to form the Independent Petroleum Association of America (IPAA) which, like API, remains very active today.

Elizabeth Mann Borghese, who was part of the Center for the Study of Democratic Institutions (founded by Bob Hutchins, my old law school dean at Yale) often argued that the oil industry was a great conspiracy. In her mind we all got together at API meetings to fix prices and set quotas. I have been a director of the API for decades and can assure you that the conflict of interest between API's many members was and is insurmountable. If you put them in a room together, they could not agree on whether or not the sun would come up the next morning. These same members have charged and countercharged each other about bad conduct in the marketplace. This just meant they were competing with one another, although they never admitted that competition was the source of their bickering.

I became a director of API in 1945, when I was president of Ashland Oil and Refining. I inherited the directorship as a representative of one of the principal independents in the country. I was reasonably active in the API for quite a few years, and I served one term as head of the marketing division. It was a great position, where I got to know most of the prominent independent marketers in the business. Since Ashland was a big supplier, my position served Ashland's interests well.

I am still a director of API. I don't know why they have not asked me to resign; perhaps the title is honorary. I don't know if I contribute much anymore, but I enjoy it and go to the annual meetings if I can. I still hear the same old squabbles as I did forty years ago—too much competition caused by the other fellow.

An Oilwoman

My oil story has been male-dominated, and you can readily see this gender bias in my references to various "oilmen." This may not be "politically correct," but it is, on the whole, an accurate representation of the oil industry. This has been a male-dominated industry throughout my lifetime, though there was a rare exception—Bettye Bohanon, who, from 1961 until her death in 1991, was Bettye B. Marshall.

I met Bettye in 1933, when she was the executive secretary of the Central Committee of California Oil Producers. I immediately

An oil woman, Bettye Bohanon, who later would be Bettye B. Marshall.

recognized her as a rare breed; a female who could go toe-to-toe with a fiery lot of oilmen. When I began my second term of servitude in Washington in 1941, it was not long before I made a cross-country phone call to her with the words, "I need help." She resisted by explaining that she was well-situated in California, and there were many things to do there. I anticipated this much and was ready with my reasons. "Bettye," I said, "you know the California oil community. Here in Washington, you will get to know the national oil fraternity, which will be invaluable later on. Besides, the country needs you. This is the most important thing you can do, professionally or otherwise." She finally agreed and left Los Angeles for Washington.

I had good reason to want Bettye on my team. She was very good with details and understood legal jargon and the engineering concepts necessary to really understand the multifaceted oil industry. She was also a good judge of people. Put these qualities together, and you have executive talent.

From 1941 to 1944, Bettye's title was Priority Specialist at the Office of Petroleum Coordinator/Petroleum Administration for War. She checked all the rules and regulations that originated from the OPC/PAW for "legal form," which meant making sure that our mandates did not even hint at favoritism or impropriety. If they did, which was rare, she brought them back to me for a review. There never was any scandal when oilmen ran the oil industry, thanks in part to Bettye.

It was after her government days that she picked up the nickname "Tiger." In the early 1960s, we were in Peru with an executive of the Texas Company (now Texaco), Ray Pittman, to evaluate a production project. In the heat of a discussion, Ray exclaimed, "You're a tiger, Bettye." I laughed and said: "You're right, Ray, and I think we have a nickname." I kept using it until it stuck, and she will always be "Tiger" to me.

After the war Bettye became executive secretary of the Conorado Petroleum Company, an exploration company drilling overseas that was headquartered in New York City. When we needed someone to be secretary-treasurer of the Minnesota Pipeline, which linked the Saskatchewan reserves in Canada to the Great Northern refining in St. Paul, Minnesota, I again persuaded her to come my way. She retired from Minnesota in 1961, when we married. A better oilwoman I never knew.

Sixty Years in Oil: Some Final Thoughts

I have been involved in the petroleum industry in one capacity or another since an evening in 1930, when, as case editor of the *Yale Law Journal*, I stumbled upon a case concerning the constitutionality of oil proration. What have I learned about the industry in the last sixty years?

I've discovered many things that I never dreamed of as a young liberal. The oil industry is not at all as the public or the media often perceive it. They regard the oil business as inherently monopolistic, though nothing could be further from the truth. The problems of the oil business do not relate to the absence of competition; many problems have resulted more from too much competition. Some of us learned that too much competition can be as bad or worse than too little. The worst of it starts with the law of capture, as interpreted in the 1860s, when a Pennsylvania judge decided there was no legal remedy to drainage.

Prodded by the observation of the physical waste of a depletable resource, I started out as an ardent conservationist. Norman Meyers and I, in our 1931 *Yale Law Journal* article, spoke of the new respectability of conservation law to arrest the waste of our depleting oil supplies. I was convinced by much reading in college and at Yale Law School that we were going to run out of petroleum. After spending most of my life fighting a surplus, however, I've come to believe it will never run out. I've had many a good argument with some of my producer friends, who say that the price of oil has to go up because increasing consumption leaves less supply. To such claims I've responded: "It is not going to run out. There will always be some new development, something we do not know about or do not expect, to increase supply faster than it is consumed."

I had one such argument with Everette DeGolyer just after World War II. He insisted that we were going to be desperately short of oil by 1956. I remember saying: "De, that prediction has been made almost since the beginning of the oil industry. Are you sure that you are right this time?" He said he was sure, and I suggested we wait until 1956 and find out. In that year, we had oil running out of our ears, and I said, "De, what happened?" He responded that some unforseen things occurred, and I remember asking, "Isn't it always that way?" I have come to the conclusion it *is* always that way.

One recent, unexpected development is horizontal drilling. None of us ever really dreamed of it as a possible technique, but it has come along nevertheless. We get oil out of the Austin chalk, which we knew was there but didn't know how to get to, by crossing the fault lines. It is a beautiful, light-gravity crude oil, and Koch Industries buys a good deal of it for its Corpus Christi refinery. Another recent example is three-dimensional (3-D) seismolog, which enhances reservoir imaging to increase the total recovery of oil and gas. Not only are explorers finding more hydrocarbons, they are drilling less wells to do it.

Another thing I have learned in my tenure in the oil industry concerns the futility of price regulation and government direction of the industry. I was writing what I thought were learned articles at Yale about how price floors and capital allocation by the government were necessary to arrest overproduction and waste in petroleum extraction. Norman Meyers and I added in our 1933 *Yale Law Journal* article that such proposals would bother "professors of 'free enterprise.'" I continued to hold these views until the secretary of the Interior asked me to draft a price-fixing order in 1933 pursuant to the National Industrial Recovery Act. My instructive, as mentioned in chapter 2, was to set minimum prices for all crude oil and refined petroleum products at every point in the United States. It was here that reality set in. I tried to draft such an order but gave up, citing the impossibility of the task. The secretary and I decided that, if we could control illegal oil production at the source, prices would take care of themselves, and we would not be burdened with trying to fix them. It was a wise decision, and it worked. Out of that decision came the Connally Hot Oil Act, which prohibited the movement in interstate commerce of crude oil or the products thereof produced in violation of state conservation laws.

As proof of the proposition that you cannot fix oil prices, a Harvard Ph.D. named James Schlesinger and his staff at the Department of Energy tried it and ended up putting the country in the gasoline lines. I once told the energy czar in person how I backed out of my chance to be a price dictator, but he had other opinions. I guess if I ever had any doubt of it before, Dr. Schlesinger proved the futility of price controls for me.

The age-old siren call by some economists to break up the "monopolistic" oil industry is as empty a public policy as has ever been pronounced. As I argued in a 1948 book review of Eugene Rostow's *A National Policy for the Oil Industry,* dismembering the major oil companies' vertically integrated operations would reduce efficiency and create a "middle man's paradise" at the consumer's expense. I still remember my analogy: breaking up the majors would be akin to chopping up the assembly of automobiles into a series of separate and uncoordinated steps. Rostow failed to understand that majors became majors by lowering costs and cutting prices and would be replaced to the extent that they failed to similarly compete in the future. There are many able independents that are always poised to take whatever the majors give them. Rostow's thesis is exactly wrong: the historic ills of the petroleum industry have resulted from too much competition, not too little.

One of the things that has always puzzled me is the line of demarcation between independents and majors. In my government days, I used to say an independent was somebody in the business whose capitalization was no more than a $100 million. Today, I would have to raise that figure substantially. Basically, the majors of the old days were integrated companies. In other words, they had production, refining, marketing, and transportation. If there is any sharp division between majors and independents, most of us would be at a loss to define it exactly. In a very general way, majors are usually engaged in some kind of marketing, while partly integrated firms are often not. But the same old fight goes on today, where the independents accuse the majors of trying to put the independents out of business by, of all things, cutting prices. I once wrote a pamphlet codifying all the laws that sought to restrain competition in one form or another. The big one was the Robinson-Patman Act of 1936, which, more or less, said that price cutting may be illegal if smaller competitors are hurt. The independents always invoked the Robinson-Patman Act, and the majors thought up ways around it. This political rivalry between majors and independents—or, more precisely, between integrated and nonintegrated firms—seems to be one of the few constants I've noticed in the oil business during my sixty years.

I have no doubts about the motivations of oil companies—big and small—that run to the state or federal government for this or that. As stated in the 1931 article Meyers and I wrote: "The driving force in the industry for stabilization has not been the zeal of the exploiters of oil to conserve supplies of petroleum for future generations but the desire to conserve their fortunes for their descendants." My experience and observations in the oil industry in the ensuing six decades have only reinforced this maxim time and time again. Pecuniary considerations come first and philosophical considerations last.

With the publication of this book, perhaps I have proven once again that I am still interested in intellectual pursuits. One of the tests of the success of such pursuits is whether or not the pursuers master their business and can interpret the wider meaning of doing so. I think I have done that. At least I've worked in every phase of the petroleum industry, from exploration and production to transportation, refining, and marketing. I have had fun from day one. Even my setbacks worked to my advantage by leading me to new opportunities and high adventures in oil. But my success is now for history to judge.

Final Letter to Ralph K. Davies

Dear Ralph,

To misquote an old maxim, "It is better to have lived and lost than never to have lived at all." I have known no one who has lived more, both in victory and defeat, than Ralph Davies. Even defeat is part of living. Certain it is that you can never know the joy of victory unless you have experienced the agony of defeat. Surely life and death follow this same pattern.

What lies ahead of what seems to be defeat of death, none of us is given to see. It needs must be something. But you, more than any of my most treasured friends and associates, have lived a great life, done great deeds, and despite the slings and arrows of outrageous fortune, won far more battles than you have lost. You have put your own particular mark not only upon the world around you, but upon those few of us who have been privileged really to know you. I count myself one of those few. I want you to know from me that you have been one of the very few who have profoundly influenced the course of my life. It would take a volume to set down innumerable illustrations. One you may not remember. In the days when we fought the battles of Sacramento you remarked, "We have a problem and we don't know what to do about it, so let's go where the problem is and we will probably think of something." We went and did think of something. I cannot count the times I have followed that fundamental rule, always recalling from whom I first learned it.

On balance, old friend, you have undone the ungodly, befriended your friends, and left the world a better place for having lived in it. I can only hope that when it comes my turn to check in, someone can say half as much of me. I would come from the ends of the earth to say all of this—and more—in person, but I am told you would rather not have even those of us who have known you best descend upon you. So be it. I know you well enough to understand the whys and wherefores.

Another of your old associates, who still laughs when she remembers how you pulled her up short for spilling "ink" (from the House of Magic) on the letter she had prepared for you, sends her love and, being a woman, a few tears. For myself, it is with a heavy heart that I write what we both know will be my last salute to you, at least on this earth. But the joys of past associations, of battles won and lost—mostly won—of high endeavors and dark conspiracies devoted to worthwhile ends, of fun and laughter—these far outweigh the anguish of the moment.

And so, until we meet again—in some brave new world where once again we may "strive, to seek, to find and not to yield."

As always,
Howard

Selected Readings

American Bar Association. *Legal History of Conservation of Oil and Gas: A Symposium.* Baltimore: Lord Baltimore Press, 1939.
Associates, Friends, and Family. *Ralph K. Davies: As We Knew Him.* San Francisco: Weiss Printing Company, 1976.
Bradley, Robert L., Jr. *Oil, Gas, and Government: The U.S. Experience.* Lanham: University Press of America, 1995.
Clark, James, and Michel Halbouty. *The Last Boom.* New York: Random House, 1972.
Frey, John W. and H. Chandler Ide. *A History of the Petroleum Administration for War: 1941–45.* Washington, D.C.: Government Printing Office, 1946.
Hardwicke, Robert. "The Rule of Capture and Its Implication as Applied to Oil and Gas." *Texas Law Review* (June, 1935)
Ickes, Harold. *The Secret Diary of Harold L. Ickes: The First Thousand Days.* New York: Simon and Schuster, 1954.
Jones, Charles S. *From the Rio Grande to the Arctic: Story of the Richfield Oil Corporation.* Norman: University of Oklahoma Press, 1972.
Marshall, J. Howard, II. "Book Review: Eugene Rostow, A National Policy for the Oil Industry." *Yale Law Journal* (June, 1948)
Marshall, J. Howard, II, and Norman L. Meyers. "Legal Planning of Petroleum Production." *Yale Law Journal* (November, 1931)

Marshall, J. Howard, II, and Norman L. Meyers. "Legal Planning of Petroleum Production: Two Years of Proration." *Yale Law Journal* (February, 1933)

Scott, Otto J. *The Exception: The Story of Ashland Oil & Refining Company.* New York: McGraw-Hill, 1968.

Scott, Otto J. *The Professional: A Biography of J. B. Saunders.* New York: Atheneum, 1976.

Tompkins, Walker A. *Little Giant of Signal Hill.* Englewood Cliffs, N.J.: Prentice-Hall, 1964.

White, Gerald T. *Formative Years in the Far West.* New York: Appleton-Century-Crofts, 1962.

Index

Adams, K. S. "Boots," 131, 189
Adelman, M. A., 41
Aetna Oil Company, 182
Agency and Refiners' Agreements, 63
alkylation, 137
Allen, Bob, 52
Allied Chemical Corporation, xiii, 8, 241–45, 248, 259
Allied Commission on Reparations, 7, 165–70
Allied Oil Company, 181–83, 189, 194
Allied-Signal Company: *See* Signal Oil & Gas Company
Amerada Petroleum Company, 52–53, 105–106
American Federation of Labor (A. F. of L.), 188
American Independent Oil Company (Aminoil), 189–93, 209, 222, 235
American Petroleum Institute, 9, 207, 260–61
Anderson, Hugo, 206
Anderson-Pritchard Oil Company, 234–35, 240
antitrust laws, xiii, 74–75, 79, 104, 266
Aramco, 111

Argentina, 8, 84, 193, 210, 211, 218–20, 229, 243
Argentina National Oil, 193
Arizona, 75
Arkansas, 40
Arnold, Thurman, xvi, 68, 79, 82, 118
Arnot, Charlie, 65, 72
Asamera, Ltd., 243
Ashland Oil & Refining Company, xiii, 158–63, 247; American Independent Oil Company (Aminoil) and, 189–92; Big Inch and Little Big Inch pipelines and, 184–85; cat cracking and, 177–80; Marshall and, xviii, 8, 132, 164–65, 170–78, 193–96, 199, 201, 202, 205, 207, 220, 222, 223, 227, 230, 234, 243, 246, 261; mergers and, 181–83; P. Blazer and, 7, 157; unions and, 186–88
as-Subah, Ahmed Al-Jabir, 190
Atkinson Proration Bill of 1939, 6, 100
Atlantic-Richfield (ARCO), 226
Atlas Pipeline, 46
Aubert, "Red," 227
Auroro Gasoline Company, 159
Austral Oil Company, 229, 236

INDEX

Bankline Oil Company, 226
Barneson, H. J., 227
Barnsdale Oil Company, 55
Becker, A. G., 207
Benedum, Mike, 37, 38
Berg, William, 85, 88, 92–93, 97–98, 108–10
Bering, Robert, 256
big dump service stations, 55–56, 73, 75
Big Inch Gas, 185, 204
Big Inch Oil, 204
Big Inch Pipeline, 6, 10–11, 128–32, 184–86
Black, John, 96
Blazer, Paul, 7; Ashland Oil & Refining Company and, xiii, 159–60, 162, 164, 170–74, 180, 182, 183–84, 188, 192, 247; Marshall and, 8, 66, 157, 158, 165, 166, 193–96, 197, 199, 202, 207, 230
Blazer, Rex, xiii, 194
Bohanon, Bettye, xix, 54, 127, 261–63
Boiler Makers Unions, 102
Boise, Idaho, 58
Bolivia, 242
Bolsa Chica Gun Club Lease, 6, 97–99, 217
Borah, William, 59
Borghese, Elizabeth Mann, 261
Bradley, Robert L., Jr., xix
Breuil, Jim, 182, 194
Breum, David, 18
Brook, Tom, 243
Brooks, Jim, 189–91
Brown, Bruce, 140, 152
Brown & Root, 185, 221
Bryant, Randolph, 47, 49, 60
Buck, E. O., 257, 258
Buck, Frank, 93, 104
Burden, William, 237
Bureau of Mines, 22
Burgan oil field, 189–92
Burrow, Harold, 258

California, 87–94, 144; Bolsa Chica Gun Club Lease and, 97–98; gasoline buying pools and, 72–79; Long Beach Oil Development Company (LBOD) and, 82–86; Long Beach Turning basin and, 95–96; Los Angeles Basin and, 20, 21–22, 216–18; Petroleum Administration for War (PAW) and, 142, 154; petroleum industry and, xi, xiii, xvii, 3, 42, 102–107, 125, 132, 163, 200–201, 216, 247, 263; regulation in, 5–7, 39, 50–57, 62–63, 99–101, 126; section 36 in, 32–34
California State Lands Act of 1938, 6. *See also* State Lands Act of 1938
Canada, 242, 263
Cardoza, Benjamin, 60
Carpenters Union, 102
Carter, Jimmy, 171
Carthaus, William J. "Bill," 207, 223–24
Catalina Island, 78
Central Committee of California Oil Producers, 51, 54–55, 62–63, 127, 227, 261; gasoline buying pools and, 74–75, 79; Marshall and, 125; regulation and, 99, 142
Central Pipeline Company, 182
Champlin Refining Company v. Oklahoma Corporation Commission, 25
Chapman, Oscar, 145
Chase, Stuart, xiv–xv
Chenery, Chris, 204–206
Chevron: *See* Standard Oil of California
Churchill, Winston, 167, 169
Clarity, Bruce, 219
Clark, Charlie, 68
Claymore oil field, 243
Coastal Corporation, The, xix, 9, 250, 257–60
Code of Fair Competition for the Petroleum Industry, 4, 28, 34–36, 51, 70, 80, 153
Cole, William, 132
Coleman, Stuart, 151–52
Cole Pipeline Act, 132
Colley, R. H., 124
Collier, Harry D., 107–108, 110–11, 113, 156–58, 182
Congress of Industrial Organizations (CIO), 188–89
Connally Hot Oil Act, 5, 41, 61–63, 265
Conner, John T. "Jack," 244–45
Conoco, xv
Conorado Petroleum Company, 263
Conroe oil field, 39
conservation, environmental, 21, 50, 53, 57, 102, 135
Continental Oil Company, xv

INDEX

Controlled Materials Plan (CMP), 127
Corcoran, Thomas, 215
Craig Shipyard, 82

Dasone, Sylvester, 206
Davies, Ralph K., xv, xvii, xviii, 96, 104, 189–90, 192–93, 229; Bolsa Chica Gun Club Lease and, 98–99; Fair Practices Association and, 75, 77–78; Huntington Beach and, 88, 89, 92–93; letter to, 269–70; Marshall and, xv, xvii, 6, 36, 53, 65, 67, 68–69, 119, 155, 184, 269–70; Petroleum Administration for War (PAW) and, 128, 145, 146, 147, 151, 154; regulation and, 62, 102; Standard Oil of California and, 108–12, 113, 156–58, 197
Defense Plant Corporation, 128–29, 137–38, 160, 162, 179, 180, 184
Defense Supplies Corporation, 124, 128, 137–38, 160
DeGolyer, Everette, 72, 97, 118, 186, 189, 264; Petroleum Administration for War (PAW) and, 122, 124, 125–126, 140, 141–42
Department of Energy, U.S.: bureaucracy of, 119, 220; regulation and, 10, 11, 35, 38, 42, 45, 140, 141, 146, 149, 171, 228, 265
Department of Interior, U.S., 4, 6, 7, 29, 32–33, 57, 116, 145, 146; California and, 50–56, 68; Code of Fair Competition for the Petroleum Industry and, 36; hot oil and, 42–49, 61; Idaho and, 58–60; Marshall and, xvii, 5, 28, 30, 31, 37–41, 64–66, 67; National Industrial Recovery Act (NRA) and, 34–35, 62–63
Department of Justice, U.S., 60, 62, 241
Department of the Navy, U.S., 165
Department of State, U.S., 165–170
Dodge, Will, 65
Doherty, Henry, 65
Dominguez oil field, 20
Donnell, Otto, 65
Donovan, William "Wild Bill," 64, 79–82
Donovan, Leisure, Newton & Lombard, 79–82
Douglas, William O., 64, 82, 238
Dunaway, Dan, 193

Eastern States Petroleum & Chemical Corporation, 220–23, 224
Eastern States Refining, 45
East Kalimantan, 243
East Texas oil field, 4, 26, 29, 37, 42–47, 49, 58, 59, 60
Eat on Industry (EOI) Club, 118
Einstein, Albert, 124
Eisenberger, Nate, 190
Eisenhower, Dwight D., 169
Elk Hills oil field, 32–34, 51
Elliott, John, 89, 90
Elmer, Bill, 239
El Segundo refinery, 22, 102, 103
Engle, Frank, 237
Esquire, 56
Exception, The, xiii–xiv, 164, 194, 247
Exxon, 243

Fahy, Charlie, 49
Fairchild, Hollis, 78
Fair Practices Association, 75–79
Farish, Bill, 65
Fascinating Oil Business, This, xi
Federal Bureau of Investigation (FBI), 58
Federal Energy Regulatory Commission (FERC), 119
Federal Oil Conservation Board, 22
Federal Power Commission (FPC), 72, 186, 238–40
Federal Register, 121
Federal Tender Board, 60–63, 84
Federal Trade Commission, 241
Fischer, F. W. "Big Fish," 5, 47–49, 59–61, 80–81, 87
Fisher, Max, 159, 199, 215, 221, 223
Fisk, Kirby, 244
Fleming, Harvey, 100
Fletcher Oil Company, 58–59, 61
Flynn, John, xv
Friends of the Earth, 102
Ford Motor Company, 82
Fortas, Abe, 7, 145, 146
Fortress of Gladewater, 47
Franklin, Wirt, 65
Frasch, Herman, 231–32
Friedman, Milton, 41
Frondizi, Arturo, 213
Frontier Refining Company, 182, 194, 228

275

Gaffar, Essat, 191
gasoline buying pools, 72–79
General Petroleum Corporation, 33–34, 56, 74, 82, 100, 227
Geological Survey, U.S., 22, 52
geology, 19
George School, 4, 15–16, 17
Germantown, Pennsylvania, 12
Germany, 167–69
Getty, J. Paul, 190, 237
Giard, George, 250
Globe Oil & Refining Company, 189, 247
Golden State Company, 103–105
Golden V, 103–105
Great Depression, 14
Great Lakes Pipeline System, 131
Great Northern Oil Company, 180, 252, 254; Marshall and, 8, 202–208, 219, 223–25, 251, 253
Green, Russ, 189, 209, 222, 228, 230
Green, Russ, Jr., 226
Guido, Jose Maria, 213
Gulbenkian, C. S., 80
Gulf Oil Company, 201

Hamilton, Walton Hale, 10, 27, 42
Hancock Oil Company, 84–85, 89, 100, 189, 208–29
Harding, Charles, 237
Hardwicke, Robert, 21, 41, 154
Harriman, Averell, 168–69
Hartley, Fred, 254
Haverford College, 4, 16–19, 23, 49
Hill, George, 125–26, 150
Hillcrest Country Club, 216–18
History of the Petroleum Administration for War, A, xii, 69
Hoffa, Jimmy, 223
Holiday, Bill, 65
Holland, Henry, 210, 212, 213, 243
Holmes, Oliver Wendell, 120
Hoover, Herbert, 101
hot oil, 5, 61–63, 73, 75, 80; in Texas, 3, 37, 42–47, 49, 58–60
Houdry Process, 135, 140
Houston, Texas, 231, 236, 238, 240
Houston Oil Company, 125–26
Howard, Frank, 134
Huffington, Inc., 243
Hulbert, Robert, 70–71

Humble Oil Company. *See* Standard Oil of New Jersey
Humphrey, Hubert, 213
Huntington Beach oil field, 20, 51, 52, 95, 97–98, 216, 225; State Lands Act of 1938 and, 87–88, 91–92
Husky Oil Company, 228
Hutchins, Robert, 23, 30, 261

Ickes, Harold, 114–17, 119, 124, 130, 158; Marshall and, 5, 28–30, 44, 66, 67, 153-155, 185; Petroleum Administration for War (PAW) and, 6, 139, 145–46, 148, 151, 152; regulation and, 4, 33, 34, 36, 37, 39, 52, 59, 101
Idaho, 58–60
Ide, H. Chandler, 69–70, 122
Illinois, 6, 125–26, 154
Illinois Basin, 126, 163, 165, 172, 174, 175, 182
Imperial Oil Company, 202–203
independent oil companies, 22, 45, 63, 73, 177–78, 201, 260–61, 266; manufacturing and, 199; Middle East and, 190; Petroleum Administration for War (PAW) and, 138, 150, 179; regulation and, 100–101; State Lands Act of 1938 and, 88–89, 90
Independent Petroleum and Consumers Association, 101
Independent Petroleum Association of America (IPAA), 261
Independent Refiners Association, 75
Independent Refining Corporation, 246–48
Indiana, 163
Indonesia, 84, 243
Inglewood oil field, 51
Ingram, Hank, 253
integration, xv–xvi
Interstate Compact to Conserve Oil and Gas, 41
Interstate Oil Compact Commission, 57
Interstate Transportation of Petroleum Product Act. *See* Connally Hot Oil Act
Isaacs, Willard, 216–17

Jeffers, William, 133–34
Jennings, Brewster, 205

INDEX

Johnson, Hugh "Iron Pants," 34
Jones, Charlie, 56, 65, 84, 144, 217, 258
Jones, Jesse, 124, 128–29, 137, 138
Jones, W. Alton (Pete), 128, 156–57

Kahle, Dick, 45–46, 220, 222
Kansas, 40, 50
Katy gas condensate field, 125
Kentucky, 159–60, 162–63, 165, 175–77
Kerr-McGee Corporation, 219
Kettleman Hills oil field, 39, 51, 52, 53, 132
Kingsbury, Kenneth R., 65, 67, 68–69, 94; Standard Oil of California and, 107, 109–112
Kingwood Oil Company, 182
Knowlton, Don, 150
Koch, Charles, xix, 9, 252–56
Koch, David, 255
Koch, Fred, 251–52, 254
Koch, William, 255, 256
Koch Industries, Inc., xix, 259, 265; Marshall and, 8, 9, 250–57
Krug, Julius, 184
Kuwait, 189–90, 243

Lake Arthur South oil field, 232–33, 235, 237, 239
Lake Maracaibo, Venezuela, 208, 242–43
Lamson, Oliver, 96
law of capture, 4, 10, 20, 21, 27, 50, 53, 73, 163, 217
Leach, Joe, 47
"Legal Planning of Petroleum Production," 4, 26
"Legal Planning of Petroleum Production: Two Years of Proration," 4, 28
Lion Oil Company, 242
Little Big Inch Pipeline, 6, 10–11, 184–86; building of, 128–32
Little Giant of Signal Hill, 199
Loeb, Henry, 213
Loeb, John, 214, 237
Loeb Rhodes investment bankers, 210, 212, 213, 219, 229, 236, 248
Lombardi, M. E. "Tex," 53, 86, 95
Long, Frank, 73
Long Beach, California, 82–87
Long Beach Oil Development Company (LBOD), 82–87, 210–11
Long Beach Turning Basin, 95–96, 99
Longview, Texas, 128
Lortscher, Frank "Dutch," 220
Los Angeles Basin, 8, 20, 55, 56, 82, 86, 216–18
Louisiana, 40, 50, 125, 176, 249
Lo-Vaca Gas Gathering Company, 258
Lycam, Donald, 208
Lyon Oil Company, 247
Lyons, Richard T. "Dick," 237

McCane, Thomas, 237
McCarthy, Glenn, 247
McCollum, L. F., xv
McDonald, Don, 229
Machris, George, 52, 247
McKee, H. Harper, 237
McKellar, Tom, 127–28
McKloskey Lime, 126
Madison Oil case, 64, 73, 79–82, 87, 115, 121
Majewski, Barney, 35, 65, 72, 131, 189, 193, 202
Mandatory Oil Import Program, 247
Manhattan Project, 145
Marathon Oil, 199
March, Harry, 89
March, O. W. "Bud," 227, 228
Markel, Frank, 232, 237
Marlago oil field, 242
Marshall, Annabelle (Thompson), 12
Marshall, J. Howard, II, xi, xiv–xvi, 4–6, 20–22, 39, 264–67; American Independent Oil Company (Aminoil) and, 189–92; American Petroleum Institute (API) and, 260; Ashland Oil & Refining Company and, xiii, 8, 164, 170–85; B. Bohanon and, 261–63; Big Inch and Little Big Inch Pipelines and, 128–32; Bolsa Chica Gun Club Lease and, 97–98; California and, 50–56; Code of Fair Competition for the Petroleum Industry and, 36; early life of, 12–18; Eastern States Petroleum and Chemical Corporation and, 220–22; Coastal Corporation and, 257–59; Fair Practices Association and, 75–78; gasoline buying pools and, 72–74; Great Northern Oil Company and, 202–207, 223–24; Hancock Oil Company and, 225–27;

277

H. Ickes and, 28; hot oil and, 42–47, 62–63; in Idaho, 58–60; Independent Refining Corporation, 246–48; influences on, xvii–xix; in law school, 23–26; Interior Department and, 29–30, 31, 64–66; Koch Industries, Inc., and, 250–56; Long Beach Oil Development Company (LBOD) and, 82–86; Long Beach Turning Basin and, 95–96; Los Angeles Basin and, 216–17; Madison Oil case and, 79–81; National Industrial Recovery Act (NRA) and, 34–35; 100-octane fuel and, 134–39, 147–48; P. Blazer and, 193–96; Petroleum Administration for War (PAW) and, xii, 113–15, 123–27, 149–55; Petroleum Development Corporation and, 249; Pillsbury, Madison & Sutro and, 94, 156–58; regulation and, 5, 7, 9–11, 37–38, 40–41, 99–102, 116, 140–46; R. K. Davies and, 269–70; section 36 and, 32–33; Signal Oil & Gas and, 8, 45, 197–201, 228–30, 241, 242, 243, 246, 250; Standard Oil of California and, 67–69, 105–107, 109–12; State Department and, 165–69; *State of Washington* v. *Standard of California* and, 70–71; State Lands Act of 1938 and, 87–94; teaching, 15, 218–19; Tender System and, 57; Union Oil & Gas Corporation of Louisiana and, 231–37; unions and, 102–104, 186–88; Union Texas Petroleum and, 238–45; W. F. Fischer and, 49, 61; World War II and, 118–22, 133
Marshall, J. Howard III, 256
Marshall, John T., 14
Marshall, Samuel Furman, 12, 14
Marshall Petroleum, Inc., 9, 250, 251
Match Play and the Spin of the Ball, 15
maximum efficiency rate (MER), 52, 90
Mayflower Hotel, 140
Meadows, Al, 177
Mendoza oil field, 212, 243
Mexico, 84
Meyers, Norman, 4, 61, 80; Interior Department and, 31, 33–35, 37–38, 40, 42, 64–65; Marshall and, 26, 27, 29, 79, 240, 264, 265, 266
Michigan, 40, 50

Middle East, 148–49, 189–93, 222; See also Kuwait *and* Saudi Arabia
Miles, J. Fred, 159
Millard, Mark, 210, 229, 236–37, 243, 244
Milligan, Bill, 253
Minkler, Bob, 123–24
Mississippi, 40
Mobil Oil, 204–06, 208, 260; See also Socony-Vacuum Exploration Company
Moffett, Jimmy, 36
Mohawk Oil, 100
Monsanto Chemical Company, 248
Moore, Ike, 252
Morgan, Frank, 219
Morgan, J. P., 256
Morgenthau, Henry, 167–68
Mosher, Edward, 258
Mosher, Samuel, 93, 95–96, 189, 220–21; Marshall and, xvii, xviii, 8, 98–99, 197, 208; Signal Oil & Gas Company and, 198, 202, 209–10, 214, 225, 229, 230, 236, 255–56
Moss, William, 249, 250
Murphy Petroleum Corporation, 242
Murray, William "Alfalfa Bill," 26

National Energy Strategy: Powerful Ideas for America, The, 146
National Industrial Recovery Act (NRA), 36, 42–47, 58; constitutionality of, 49–50, 53, 60, 63, 72; formation of, 34–35; petroleum industry and, 4–5, 37–41, 62, 64, 73, 115, 142, 265
National Labor Relations Act, 36
National Labor Relations Board, 102, 188
National Petroleum Association, 9
National Petroleum Council, 9, 184
National Policy for the Oil Industry, A, 266
National Stripper Well Association, 9, 178
Natural Gas Act of 1938, 238
Nelson, Donald, 130
Nevada, 75
Newman, Floyd, 181
New Mexico, 40, 50
New York Stock Exchange, 254–55

278

Norris City, Illinois, 128–32
Norwalk Oil Company, 219, 226

Office of Petroleum Coordinator for National Defense. *See* Petroleum Administration for War
Office of Price Administration (OPA), 142–45, 149
Office of Rubber, 133–34
Ohio, 176
Oil and Gas Division: *See* Federal Tender Board
Oil & Gas Journal, 49, 68
Oil code. *See* Code of Fair Competition for the Petroleum Industry
Oil Conservation Commission, 100
Oil Umpire's Office, 51, 74
Oil Workers Union, 99–100, 102, 165, 186–189
Oklahoma, 50, 65, 83–84, 192, 249, 252; regulation in, 4, 5, 22, 25–26, 40, 41, 51
Oklahoma City oil field, 4, 26
Olson, Culbert, 89, 101
100-octane fuel, 134–137, 160; Petroleum Administration for War (PAW) and, 138–40, 147–48, 153, 179
Oregon, 75
Organization of Petroleum Exporting Countries (OPEC), 41
O'Shaughnessy, I. A., 189, 247
Oxley, John, 235, 237, 244
Oxley, Patty, 251

Pacific Coast Petroleum Agency and Refiners' Agreement, 73, 75
Pacific Coast Petroleum Agency and Refiners' Association, 62
Pacific Electric Railroad, 87, 90
Pacific Union Club, 109
Panama Refining Co. v. Ryan et al., 49
Panhandle Eastern Corporation, 186
Papoose, 58
Parten, J. R., xvii, xviii, 8, 253; Great Northern Oil Company and, 202, 204, 205, 207, 224, 225; Petroleum Administration for War (PAW) and, 124, 128, 147, 151
Pauley, Ed, 65, 89, 90, 166–70; World War II and, 115–20
Pearson, Drew, 130, 151

Pelly, James, 132
Pemberton, J. R. (Bill), 54
Pennsylvania, 163, 176
Perino's Restaurant, 92–93, 218, 230
Perón, Juan, 210, 229
Persian Gulf, 245
Peru, 263
Petrofina, 228–29
Petroleum Administration for War (PAW), 133, 151–52, 157, 166, 184, 263; Big Inch and Little Big Inch pipelines and, 128–32; 100-octane fuel and, 134–39, 147–48; preparation for, 113–22; Marshall and, xii, 5–7, 30, 62, 153–55, 179, 251; regulation and, 125–26, 140–46, 149–50; World War II and, 123–24, 127
Petroleum Administrative Board, 50
Petroleum Conservation Commission: *See* Federal Tender Board
Petroleum Development Corporation, 249–50
Petroleum Industry War Council, 36, 114, 120, 123–24, 150
Petroleum Requirements Committee, 144
Pew, J. Howard, 11, 65, 139, 140
Pew, Joseph, 65
Philadelphia Enquirer, 19
Phillips, Frank, 131
Phillips Petroleum Company, 131, 132, 189–93, 238
Pierce, Eleanor, xix
Pillsbury, Madison & Sutro, 94, 103–104, 116, 156–58, 246
Pipe Fitters Union, 102
Piper oil field, 243
Pittman, Ray, 263
Poe, Holly, 186
Port Arthur, 46
Powers, Elliott, 243
Presidio Oil Company, 250
price-fixing. *See* regulation
price wars, 55
proration, market-demand, 4–5, 10–11
Pure Oil Company, 208, 209, 226, 253, 254
Pyles, Ernie, 86

Raisies, Consintine, 23
Ralph K. Davies: As We Knew Him, xviii, 190

Randolph, Roger, 237
Redondo Beach, 216–17
regulation, 100, 116, 140–46; in California, 50–56, 62–63, 99, 101–106; Dept. of Energy and, 10, 11, 35, 38, 42, 45, 140, 141, 146, 149, 171, 228, 265; in East Texas, 39, 42–47; in Idaho, 58–60; in Illinois, 125–26; Marshall and, 4–5, 9–11, 26–28, 31, 61, 264; National Industrial Recovery Act (NRA) and, 34–38, 40–41, 49; petroleum industry and, xii, 6–7, 22, 25, 185; Tender System and, 57; World War II and, 114
Reid, Will, 85, 86, 189, 225, 226
Republic, 58
Reynal, William "Billy," 210, 213
Richardson, Sid, 54
Richfield Oil Company, 217, 219
Rio Vista, California, 105–106, 107
Robinson-Patman Act of 1936, 75, 266
Rockefeller, John D., xiv, 3, 30, 93, 209
Rockefeller, John D., Jr., 109, 111
Rockefeller, Winthrop, 93, 96
Rock Island Oil & Refining Company, 252, 253
Roebuck, Donna, 251
Roeser, Charlie, 126
Roosevelt, Franklin Delano, 28, 101, 114–16, 119, 124, 153, 166; National Industrial Recovery Act (NRA) and, 36, 38, 44
Rostow, Eugene, 266
Rowe, Professor, 19, 20
rule of capture. *See* law of capture
Russell, Frank, 237
Ryan, Archie, 47

Sacramento, California, 105
St. Clair, Press, 54, 65
Saks, Alexander, 124
San Joaquin Valley, 227
Santa Fe Springs oil field, 20, 51
Saudi Arabia, 41, 111, 190
Save the Beaches Association, 90
Schaefer, Ben, 187–88
Schechter Poultry ("sick chicken") case, 35
Schlesinger, Henry, 250, 251
Schlesinger, James, 11, 140, 265
Schroeder, George, 106

Scoal, Dave, 220
Scott, Otto, xiii–xiv, 164, 194, 247
Scully, Bill, 75
Scurlock, Eyvonne, 251
Scurry County oil field, 201
Sean gas field, 243
Second War Powers Act, 141, 142
section 36 decision, 32–34
Securities & Exchange Commission (SEC), 254, 258
Sharkey Bill, 101
Shell Oil Company, 22, 56, 103, 111, 183, 243
Sherman Antitrust Act, 75, 79
Shoreham Hotel, 122, 126
shortages, 133–34, 170–78
Shreveport, Louisiana, 46
Sierra Club, 102
Signal Hill oil field, 20, 21–22, 51, 200, 225
Signal Oil & Gas Company, 56, 84–85, 89, 98, 100, 189; Eastern States Petroleum & Chemical Corporation and, 220–22; Great Northern Oil Company and, 202–207, 223–24; Hancock Oil Company and, 225–27; Long Beach Turning Basin and, 95–96; Los Angeles Basin and, 216–17; Marshall and, 8, 45, 197–201, 228–30, 241, 243, 246, 250; S. Mosher and, xviii, 236, 255–56; in South America, 208–16, 218–19
Sinclair, Harry, 65, 84
Sinclair Oil Company, 226
Smith, Adam, 27
Smith, Felix, 53
Smith, Warren, 258
Socony-Vacuum Exploration Company, xviii, 111. *See also* Mobil Oil
South America, 218–19; Signal Oil & Gas Company and, 208–16
Southern Natural Gas Company, 204
Southern Production Company, 204–205, 206, 252
South Saskatchewan Pipe Line, 204
Southwest Exploration Company, 89–90, 98–99
Soviet Union, 140, 166–70
Spencer, Percy, 252
Sprayberry trend, 175
Stalin, Joseph, 167

INDEX

Standard Oil of California, 21–22, 34, 56, 80, 84–85, 101, 166, 189; Atkinson Proration Bill and, 99–102; Bolsa Chica Gun Club Lease and, 97–99; Long Beach Turning Basin and, 95–96; Marshall and, xiv, 5–6, 8, 53, 67–69, 94, 102–105, 148, 157–58, 182, 228, 246, 260; presidents of, 107–12; R. K. Davies and, xv, xvii–xviii, 62, 113, 116; *State of Washington* v. *Standard of California* and, 70–72; State Lands Act of 1938 and, 87–88, 92–93
Standard Oil of New Jersey, 39, 111, 126, 134, 152, 201
Standard Oil of Ohio, 128, 172–73, 185, 186, 209
Stanley, E. N., 44
Starr, Gene, 217
State Lands Act of 1938, 6, 87, 88–90, 92–94
State of Washington v. *Standard of California*, 70–72
Sterling, John W., 30
Sterling, Ross, 26
Sterling Memorial Law School, 30
Stewart, R. McLean "Mac," 229, 230, 236–37, 239, 244
Stone, Judge Patrick, 80
Stimson, Henry, 145
Strook, Bill, 60
Submerged Lands Act of 1953, 214, 215
Suebert, Ed, 65
Sugar Hill oil field, 218
Sumatra, 243
Sun Oil Company, 135, 139, 140, 209
Superior Oil Company, 52–53, 209
Supreme Court, U.S., 64, 75, 82, 103, 104, 214–16, 238; National Industrial Recovery Act (NRA) and, 5, 53, 63, 72
Sutro, Alfred, 94
Sutro, Oscar, 34, 36, 65, 68, 112, 158
Swarthmore College, 16, 18
Swensrud, Sydney, 65, 128
Swidler, Joseph, 240
Swiss Oil Company, 159–60

Tarbell, Ida, xiv–xv
Tax Reform Act of 1986, 250
Teagle, Walter, 65
tea-kettle refineries, 55, 63

Teamsters Union, 223
Tender System, 57–63
Tennessee Gas Transmission, 242
Ten Years of California Oil, xi
Texas, 176, 222, 249, 258, 259; hot oil in, 37, 42–47, 49, 58, 59, 60; Petroleum Administration for War (PAW) and, 141, 145, 148, 150, 154; regulation in, 4, 5, 7, 26, 39, 40, 41, 50, 57, 125
Texas Commerce Bank, 9
Texas Company, 77, 148–49, 153, 189, 263; Ashland Oil & Refining Company and, 176–77; Madison Oil case and, 79–82. *See also* Aramco
Texas Eastern Transmission Company, 132, 185–86
Texas Gas Transmission Company, 238–239
Texas Monthly, 257
Texas Natural Gasoline Corporation, 235, 236, 238, 240
Texas Railroad Commission, 5, 43–46, 141, 143, 258
Texgas, 241
Thomas, J. Elmer, 37, 51
Thomas Cook & Sons, 24
Thomas-Disney bill, 57–58
Thompson, E. O., 44, 142
Tidelands. *See* Submerged Lands Act of 1953
Tidewater-Associated, 56
Tilden, Bill, 4, 15–16, 18
Tompkins, Walker, 199
town-lot drilling, 70, 73, 125, 216, 218
Truman, Harry S., 7, 115, 166–70, 184
21 Club, 229
Tyler, Texas, 60

Union Oil of California (Unocal), 52–54, 100, 218, 226, 254
Union Oil & Gas Corporation of Louisiana, 229, 231–37. *See also* Union Texas Petroleum
Union Pacific Railroad, 82
Union Texas Petroleum, 8, 238–40, 246, 247, 250, 257; Allied Chemical and, 241–45
unions, 99–100, 102–104, 180, 186–89
United States v. *The Standard Oil Company, The*, 33
unitization, 65

Universal Oil Products, 177
U.S. Steel, 12

Valvoline Oil Company, 182
Vandeveer, W. W., 181, 189, 194
Venezuela, 208–10, 242
Ventura Avenue oil field, 51
Virginia International Company, 243
Von Fleet, Bill, 63

War Production Board, 115, 119, 127, 130
waste, xiv, 4, 26
Weil, Al, xiii, xvii, xviii, 56, 110, 227, 260; gasoline buying pools and, 73–74, 75, 77; section 36 and, 33–34
Wheeler, Burton, 215
Whiteford, Bill, 98
wildcatters, 22, 33, 163, 201, 211–12
William Moss Properties, Inc., 249
Wilmington oil field, 7, 82–84, 95, 99, 145, 225
Wilshire Oil Company, 52, 247
Wilson, Robert, 115, 119
windfall profits tax, 229
Winkler, L. E., 251
Winkler-Koch Engineering, 251
Winnie, Texas, 247, 248
Winter, Art, 223–24, 226
Wisconsin, 79–82

Wolf, Justin, 127, 147
Woodley Petroleum, xviii, 204–05, 206, 253
Wood River Oil & Refining Company, 252
Wootan, Jim, 218
World War I, 21, 36, 168, 235, 260
World War II, 96, 113, 128–32, 264; Petroleum Administration for War (PAW) and, xii, 7, 69, 123–127, 140–46, 149–52, 154–55; Marshall and, 53, 116–22, 251; 100-octane gasoline and, 134–39, 147, 148, 153; petroleum industry and, 3, 6, 10–11, 36, 85, 113–15
Wyatt, Oscar, xix, 9, 257–60

Yacimientos Petroliferos Fiscales (YPF), 210–13, 219–20
Yale Law Journal, 4, 23, 25, 27, 28, 146, 264, 265
Yale School of Law, 4, 5, 23–26, 84, 264; teaching at, 27–29, 67, 68
Yalta Conference, 167–68
Yarborough, Fletcher, 258
Yorty, Sam, 100
Young, Garth, 95, 149, 190, 191, 209

Zukoff, Georgi K., 169